I HAVE ADHD/ADD – SO WHAT?

GUIDELINES FOR
TEENS AND ADULTS

Dr Marius Potgieter

Copyright laws and legal disclaimer

Dedicated to

All teens with or without a diagnosis of ADD/ADHD.

My grandchildren: Ruth (Melody), Lean, Suzanne, Jean, Catherine (Katie), Kelly, Jesse, Marius (Mario), Andre, Helinda, Tracy, Joshua, Christi, Johan, Emma, and...

And everybody with symptoms of ADHD, including myself, with a reminder to "practise what I teach".

Acknowledgements

My sincere thanks to:

At St Peter's Hospital, Chertsey: Dr Wajdi Nackasha and Dr Bozhena Zoritch, consultant paediatricians who have inspired me, and Dr Fivos Cacoullis and the other child psychiatrists and CAMHS professionals with whom I have worked closely. All my other colleagues, patients and their parents over many years. Diane Morrison for her ongoing help with the editing. Holly Alexander for her ideas, editing and support when I wanted to give up. Linda, my wife, for her constructive criticism and insurmountable patience. Andre Potgieter J.D. for his support and ideas. Mario Ebersohn for the "crocodile" illustration, and for his help with revising the teenage part. Dr Thomas Brown for giving me permission to incorporate his Executive Function model. Author-partners and Lulu Publishers for their help with editing, formatting and publishing of the book. Last but not least, my Heavenly Father for His love, mercy and protection.

A colleague of Dr Potgieter writes about the Teen section of this book:

"This book fills a much-needed place for providing support for teens with ADHD. I have found it difficult to find information out there for adolescents with ADHD and I will certainly find this a useful tool. Your clear, concise and easy-to-read language is a great plus. You've tackled many of the key problems that adolescents frequently don't want to talk about, such as relationships, and have also emphasised the positives of having ADHD. I like the way you've given people things to think about and strategies to try, and that you understand the reality of teens having ADHD and what they have to struggle with in their own individual ways."

"I highly recommend this book to teenagers with ADHD as an excellent resource for them."

**Dr G D (Geoff) Kewley, Consultant Paediatrician,
The Learning Assessment & Neurocare Centre,
Horsham, West Sussex**

About the Author

Dr Marius Potgieter is a paediatrician who has been treating teenagers and children suffering from ADHD for over 30 years.

As many of his patients grew older, he developed a keen interest in helping them take control of their ADD/ADHD during the transition to their teen years and then into adulthood. As part of that work, he discovered that the parents of many of his patients had ADD/ADHD as well, as it runs in families.

Dr Potgieter has published books for adolescents and adults to help them understand more about ADD/ADHD. With this knowledge and through taking more responsibility for their actions, they are able to adjust their symptoms to their advantage. In this way, people with ADD/ADHD are able to take care of themselves, many even without medication, especially if they receive the understanding and support of those around them. As a doctor, he also acknowledges that there are some people who need more help by way of professional medical and psychological interventions. Successful treatment must match each person's level of symptoms.

He currently works as a paediatrician in Surrey, England, where he lives with his wife who has supported him for over 45 years. He has children and grandchildren in the UK, USA and South Africa, and friends – including many of his former patients – all over the world.

Contents

THE TRANSITION YEARS

HELP FOR ADULTS

FURTHER RESOURCES

I have ADHD/ADD – so what?   Dr Marius Potgieter

www.adhd123.net

I have ADHD – so what?

Yes, Dr Marius Potgieter has ADHD. My parents had a hard time with me as a teen. They were so concerned by my behaviour that they eventually sent me to a boarding school in a small town where my brother was a teacher. Needless to say, as a "city boy", I managed to teach the other kids a thing or two about life, but when it came to schoolwork it was a different story. I always had to work much harder than many of my peers, though I was tested as having a normal IQ. I once asked my woodwork teacher; "why does it always take me much longer to finish my drawings than the other children?" His answer was: "I think you don't plan ahead."

With hard work (getting up in the mornings to study when the others were still sleeping) I finished school with a few distinctions, and high enough grades to be selected to go to medical school. Studying for me was never easy but I finished successfully, worked in a hospital for a number of years, and then became a consultant paediatrician in a private practice, and also an Associate Professor at the Medical University of Southern Africa.

I tell you all this for one reason only and that is to show what can be done with dedication, will-power and faith, never forgetting the support of the people around me. The principles that I write about in this book have worked for me personally and have helped hundreds of my patients to overcome their problems caused by ADHD.

I admit it hasn't always been easy and it wasn't until my later years that I asked my GP to refer me to a psychiatrist, who started me on medication. Though the effect tends to wear off in the late afternoon, and I may have some problems going to sleep, I don't have to "push" so hard. It keeps my thoughts "less scattered" and I'm less inclined to "doze off" in meetings. Medication should never be used as a crutch, however. The will-power to order your life according to the basic principles in this book is what will really bring about positive change.

I like the song, *"You raise me up to more than I can be…"* It makes me think both of my limitations, and my responsibility to try every day **"to be the best me I can be"**. Sometimes when I am tired or discouraged I like to read letters I received from patients. It helps me keep working on ADHD. I hope it will help you too.

Dear Dr Marius,

I looked at one of the reports you copied to my parents after we saw you, and thought just to drop you a line.

You very much know how much I struggled before I was brought to you at the 'late age' of 14 by my mum and dad. I remember that I did not want to come because I was tired of being told that I 'could do better if I only wanted to.' I felt like nobody understood me, though my parents tried hard to do so. For my sister two years younger than me, I was a continual embarrassment, but I thought she was only nasty to me all the time.

With your help and the support of all those who suddenly started to understand me, I am now in my second year at college, and though I have regular blips and hitches, they don't get me down.

Thanks again for all your help,

David Bromfield, West Byfleet, Surrey

How to use this book

This book is designed for all ages to help you overcome the problems you face with ADHD and turn around what you might have thought of as negative characteristics to your advantage.

Remember that some of the cleverest and most creative people have ADHD. The ability to lead a healthy and happy life is already inside you. All you need to do is **redirect your energy** and this book will show you simple ways to achieve this.

The first thing I would like you to do is to practise:

1.   **THINK.** Thinking is important, because this is where it all starts. But it shouldn't end there…

2.   **SAY.** Speaking closes the gap between thinking and doing. **Voice** your thoughts, **discuss** them with others and then **write them down or draw pictures.**

3.   **DO.** Doing is what it eventually boils down to, and not only doing, but also **finishing.**

This process will not only help you through all the exercises in this book but will help you in virtually every aspect of your life. You may have noticed that my website is www.adhd123.net – the 1-2-3 is a handy reminder to Think, Say, and Do.

What I want you to get out of this book is to be the "best me" possible. Let's start by doing some very simple exercises to show you how you can put Think, Say and Do into action and start working on your problems.

Right now, or when you wake up in the morning, say to yourself:

"This is the first day of the rest of my life and I am going to make it my best day!"

Get into the habit of saying this each day. Then say:

"Today I am not going to waste opportunities, time or money. I am going to develop ways of thinking, saying, and doing to become the best me I can be."

Think about what you are going to do differently today, talk about it with your friends, family or colleagues, and make notes. Ok, this might sound obvious but it's important – the "doing" part actually involves another "d". Before we can "do" something, there is a **"decision"** to make. If you decide to do something with **enthusiasm,** because you believe it's the **right thing** to do, you are much more likely to act on it. Always try to make **informed** and **firm** decisions and **write down** what you decide so that you'll remember it.

Ask yourself these two questions before deciding to do something:

❖ **What can I gain by doing this?**

❖ **What can I lose by doing this?**

These questions apply equally to the advice in this book and to the rest of your life. Maybe we can't always choose our own circumstances, but we can choose how we respond to them.

As Aristotle, the Greek philosopher, said: "We are what we repeatedly do". By reading this book, you have already taken a decision to improve your life, "to be the best me I can be". The good news is that you can do it and the advice given in this book has been proven to work for both adults and teens. But you will need to practise. That means doing things over and over until they become a habit. It's hard to start with and you will have bad days, but once you start to Think, Say and Do, life will get easier as you learn to manage and overcome the problems your ADHD may cause.

Don't think you need to tackle everything at once. Do what's most important for you. Most of the practical advice is in the section for teens and the section for adults. If you want to get straight to work, jump to one or other of these sections. The strategies for overcoming problems are the same for both teens and adults but, as you grow older and take on more responsibilities, the nature of the problems tends to change. Poor concentration and impulsiveness may stay with you as you get older, but hyperactivity is replaced by an inner and outer restlessness. Problems in the classroom and with peer relationships change later to problems in the study-place, work-place and with partners, family and friends. There is a whole section devoted to ADHD in the transition years from teen to adult.

The other major section in the book explains more about ADHD, both the positive and negative aspects. It is designed to help you understand your problems better and you may find it useful to share this information with family and friends.

This book is NOT intended to be a diagnostic tool for ADHD. If you think you might be suffering from ADHD and are struggling to overcome the problems you are experiencing, you should go and discuss it with your GP. Your doctor may decide to refer you for more specific professional help, and possibly medical treatment.

Although this book is written primarily for people with ADHD, there are others, like right-brained thinkers, who may share some of the symptoms and problems. If you have problems with restlessness, being attentive, forgetting and losing things, remembering to do things, and organising your life generally, the techniques in this book will help you too. But if you suspect that you suffer from a medical condition, such as depression, anxiety, or a stress-related illness, do consult your GP as you may need or benefit from expert help.

What is ADHD?

ADHD stands for **A**ttention **D**eficit **H**yperactivity **D**isorder. When children with ADHD are unable to concentrate, but are not hyperactive, some people call it ADD, without the "H", but most people just call it ADHD anyway.

About one in every 20 people has ADHD, so you are always around others who share some of your challenges and are not the odd one out. About one in every 10 people is left-handed, which you probably never

One in every 20 people has ADHD

notice unless you are right-handed and happen to sit next to a left-handed person when you're eating at a crowded table. Just so with ADHD – you can probably make it mostly invisible to others.

ADHD was first recognized during the mid 1800s. In 1902, a British paediatrician named George Frederic Still was probably the first to observe and describe it to the Royal College of Physicians. As far back as 1937, it was noted that certain medications might help children who suffer from inattention, hyperactivity and impulsiveness.

However, it wasn't until the 1950s and 1960s that medication like Ritalin (methylphenidate) began to be prescribed for ADHD. Today it's prescribed to millions of children and adults.

Most people with ADHD are born with it but, before it's identified, others often think that signs like restlessness and difficulty in concentrating mean that children are naughty or badly brought up. The parents are often blamed for their children's misbehaviour. Boys are more frequently diagnosed than girls when they are young because they tend to be more disruptive in the classroom but, later on, a higher percentage of girls are diagnosed.

Many people with ADHD are unusually clever or gifted

On the positive side, it is also clear that many people with ADHD are unusually clever and gifted, but they need help to channel their activities in the right direction. People with ADHD may be very good at multitasking, solving problems, and handling crises, all of which are good life and career skills. This book will help you control the negative side and bring out the positive side of ADHD.

The basic problem with ADHD centres on low levels of neurotransmitters (substances between nerve endings that help impulses go from one nerve to another) in certain areas of the brain, particularly those that control attention, activity, impulsiveness, and the ability to start and finish tasks. There are several of these neurotransmitters, the most important of which are serotonin, dopamine and nor-adrenaline.

Positive thinking and physical exercise can help you a lot

Neurotransmitter levels may vary in everyone, but the "default" level for someone with ADHD is low (less than sufficient). However, your levels are naturally increased when involved in physical exercise or when doing something particularly interesting or exciting. For example, children involved in sport often do better with their schoolwork. This is also, why you may be able to sit down for hours and play a computer game or work on a project that really motivates you, but struggle to get through a class or a work meeting that may not "grip your attention."

The increase in neurotransmitters will drop again but launching into things with determination and enthusiasm can certainly help you overcome the deficiency. Newer brain research, such as that described by Dr Caroline Leaf in her book: "Who switched off my brain?" shows that the traditional view that our brains are "hard wired" is not true. The brain circuits and even structure can be changed by consistent thought patterns (negative thoughts bring about a decline or shrinking of the nerve cells in our brains, and consistent positive thoughts bring growth and new nerve connections). It could be that some people who have "outgrown" their ADHD are the ones who have unknowingly applied this principle consistently! Others learn to cope with it, and about one-third really need ongoing medication.

Always remember that ADHD is primarily a **medical condition** with a physical cause, just like diabetes or a thyroid problem. Don't let anyone convince you that ADHD is mental – that often causes people to give up, go off their medication, or stop doing things in ways that help them cope well with ADHD, making their symptoms worse instead of better. There are environmental factors that may make the symptoms worse; like problems with relationships or difficulties at home, school or work, but they don't cause it. There are a number of other conditions that also depend on the correct level of neurotransmitters and symptoms of these conditions may overlap with those of ADHD. Some examples are depression and extreme anxiety, which the doctor who diagnoses ADHD has to keep in mind.

The reason why people with ADHD produce low levels of neurotransmitters is still unclear. All we know is that genetic factors may play an important role, so some other members of the family are also likely to have it, or at least symptoms of it.

ADHD may be present from an early age, but it often becomes more obvious when children attend school, and is most likely to be first noticed by a teacher. Diagnosis is not always easy, and is usually made by a specialist, most often a paediatrician or a psychiatrist who works with children. He or she needs as much information about the person as possible, usually from parents, teachers and the patient. Reports from other professionals may be necessary as well.

Some people are not diagnosed with ADHD until they are adults. This makes it harder for them to adapt because they have built up so many bad habits over the years. If you have not been diagnosed, the best thing to do is to ask your doctor, who may refer you to specialist practitioner who can really help with your particular problem.

Types of ADHD

There are three basic types of ADHD, which show their symptoms in different ways. There are also a number of related conditions, which further complicate the picture and make diagnosis more difficult.

Hyperactive-impulsive ADHD

Some mums notice that their baby is very active even before birth. After birth ADHD children may be more vigorous and irritable than most babies, and may suffer from stomach cramps. Sometimes they vomit as if they have problems with feeding (this may be true but it's not always easy to tell what is causing it). The child remains very active, always on the go, and may (but not always) walk sooner than most children. They may also have problems with balancing, and take longer to ride a bicycle.

They usually go from one thing to another without finishing what they were doing, and may even stay active after bedtime. When they sleep, they may still be restless, often waking up several times during the night. As they grow older, they have lots of energy, as if driven by an engine.

Later, in nursery school and in the classroom, students with ADHD can't sit still and may get up from their chair many times, becoming disruptive. Because of this continual behaviour, they may annoy others and then feel rejected if confronted by other students or their teachers. The problems often get worse before they get better, because at first everyone, including their parents, often believes that they are behaving badly on purpose. This can make people with ADHD feel rejected. They may become very angry inside, start to be rude and continually argue.

People with ADHD don't always understand exactly how they are annoying people. They can't figure out why people respond unkindly towards them, and often feel they are being treated unfairly. They then try to protect themselves by being rude as well, and they may be excluded for short periods or even expelled from school.

All of this may make it hard to make friends, and easy to become lonely. When you feel lonely, it's tempting to accept friendship from people without thinking much about what kind of people they are. You may find that some friends have quite a few troubles of their own, whether they have ADHD or other kinds of problems. It is important to have friends who are there when you need them, and to make sure that you are also there for them if they may need you. However you should not allow friends to persuade you to do what you know is wrong.

Inattentive ADHD

This is more common in girls than in boys, and the problems are initially less obvious. Children may be contented as babies but later become more restless and unable to pay attention to one thing for more than a short period. They are inclined to daydream but may also be distracted by what's going on around them. Other children may consider them "odd" and even make fun of them. The symptoms are likely to become more obvious when children start formal education. Though they may be bright, they often underachieve and may not concentrate well on schoolwork.

With the correct diagnosis and treatment, followed by learning to manage your behaviour, you can overcome the problems associated with this type of ADHD

Mixed type ADHD

This is where the symptoms of hyperactivity/impulsivity and inattentiveness are both present at the same time.

Again diagnosis, treatment and understanding can prevent many problems and unhappiness.

Related conditions

People of all ages with ADHD may also develop related conditions. One, as mentioned earlier, is extremely difficult behaviour, which is sometimes referred to as oppositional defiance or ODD. This is when you always know better than anyone else, oppose what your parents, teachers, or boss want you to do, often do things they tell you not to do, and spend a lot of energy arguing or doing things simply because you think it will upset them.

Sometimes you do things you would prefer not to do when you think about them, because being defiant becomes a habit. Many times you may feel you are right when you oppose others. This may sometimes be true, but remember you can't be right and all the others wrong all the time.

ODD is not an integral part of ADHD and can occur without it, but statistics suggest that it is present in more than 60% of ADHD cases. It is generally very obvious in teenagers but can be quite subtle in adults. Multiple factors may contribute to the development of ODD and there are numerous articles and books available on the subject. My opinion is that it has to do with defending your self-image and a frustrated need for self-expression. Secondary gain may seem like the overwhelming reason for the person's behaviour but, in this case, it's not the primary cause.

People with ODD may feel others are not listening to them. It also really hurts if someone "interprets their motives" for them. Rightly or wrongly they may think the intention behind their action was reasonable and feel that the person who tells them off about it doesn't understand them. The worst thing someone in authority can do is to interpret the motives of someone with ODD. Even though they may accept being told off for doing something wrong, they don't need to be told, "You did this because..." There is a saying: *"judge the act, not the person"*. It is unfair to judge someone's intention: only that person can know it and, even then, they may not really understand it. Ask them their reason for saying or doing something and listen actively to what they say.

Probably the most important thing with ODD is to find opportunities for healthy self-expression through activities and relationships. (I like the meaning Timothy Ferriss gives to ADD in this regard, as an "Adventure Deficit Disorder".)

Consistent and reasonable boundaries maintained by a sympathetic authority figure will help the person with ODD to "internalise" healthy behaviour patterns and to come to emotional maturity, so there will be less need for "external" demands and constraints.

Just like ADHD, ODD can be a very complex condition and may be associated with other problems including learning disabilities, epilepsy, and autistic spectrum disorders, amongst others. Martin L. Kutscher's book, *"Kids in the Syndrome Mix"* is an excellent book on how these conditions can overlap.

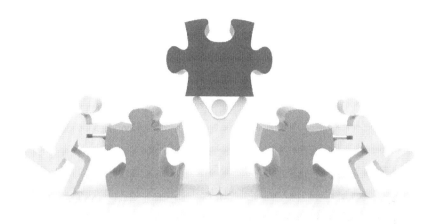

HELP FOR TEENAGERS
Getting started

Virtually every teenager is more emotional just before their early teenage years. Does hearing phrases like: "You never used to be so stressy..." "Why don't you ever do your homework?" "Stop being so mean to your sister!" or "You keep being stupid... " come to mind now? The ability to think straight will come back sometime later on in your teenage nightmare. To complicate things further, young teenagers find it difficult to relate to other people, so they might be unhelpful, appear uncaring and be misjudged by their friends, parents and sometimes even strangers. This changes gradually. Remember,

You are never too young to start taking responsibility for yourself.

Think about it. You are responsible for your actions, for your choice of friends and for deciding which people you choose to pay attention to. These choices will determine what you think, and what you think will determine what you do and, in the end, who you become.

You have three things to spend – your time, your money and your energy. The more sensibly you spend them, the more they will become "available". Try it! Determine your own goals, and use your **"brain energy"** to chase them. Get in the habit of doing today's tasks today.

Some of the problems you might be experiencing you will lose, others you will be able consciously to overcome. There might be some problems that will stay with you. This book is to help you overcome those problems. You can develop coping strategies so problems don't overcome you. **You do possess the strength to overcome the difficulties in your life and, when necessary, to accept help from the right people. You will also develop the strength to ask for help when you need it.**

The transition from childhood to adulthood doesn't happen overnight. You might start thinking like an adult at different ages and as different experiences happen in your life. "Thinking like an adult" is not boring. It's cool, and it means you are taking responsibility and choosing the path for your own life. Sometimes your mates and your family may be thinking the same as you, sometimes they might seem to be holding you back or pushing you in a direction you don't want to go.

A star won't let bad songwriters or musicians take over their band. In the same way, you need to be the leader in managing your life so it doesn't become a long sad song. Some people with ADHD use it as an excuse for doing bad, stupid and destructive things. When you understand yourself better, you can prevent the things that might go wrong and let your creativity shine through.

The sooner you digest the tips contained in this book, the better the outcome will be for you. Use it as a companion on your life's journey.

You may not have given it a thought, but of the approximately six billion people on earth, there is nobody else exactly like you. Your eyes, your ears, your brain and everything else about you, even your fingerprints, are your very own.

Think what will happen if you use everything you have to the best of your ability...

Tell yourself "**I am going to be the best me I can be.**" Not to be better than others, but the best YOU can be yourself. You can always find someone who is better or worse than you at some things but they will never be a better you. Don't compare yourself with others; it's a big waste of time.

The following habits have helped people with ADHD be their "best me". You may want to make them part of your life:

1. **Understand yourself and the things around you.** Don't be afraid to develop your own opinion. Read, listen, and ask questions. Get answers about yourself, about ADHD, and about any other problems that might be bothering you.

2. **Look around you and observe.** Watch how people respond to different things. Make it a game to guess what people are going to say before they say it. You won't always be right, but it'll be fun. One problem people with ADHD tend to have is being impulsive. Developing the

ability to predict how others may respond will help you stop and think before you act impulsively.

3. **Listen and hear what is going on around you.** Listen, listen and listen! Listen to what others say. Listening is a skill that needs to be learned through practise. You'll meet a lot of people who are not very good at listening.

4. **Think about what you see and hear** and try connecting it to what you already know.

5. **When it is your turn to speak, speak sensibly and speak clearly.** Don't be afraid to have your say. You won't always be right, but no one is always right. Give others a chance to speak and pay attention to what they have got to say.

6. **Action.** All the other things come to this – to **do!** Do everything to the best of your ability, and never say you are going to do something and then not do it. Forget the phrases "I can't do it" and "I am bored."

7. **Organise.** You must have a basic plan for organising your life and your things.

This last point is very important, because staying organised will always be a challenge for you, and you probably have already learned that being disorganised creates problems. If not, someone else is keeping you organised and it's time for you to take over that job yourself.

Make a list of things you have to do every day and write "done" over every task you complete. Concentrate on one thing at a time, but keep in mind what you are going to do next. Decide to be happy and excited when you do things. It makes them a lot easier. Even if the things themselves may not be so exciting, look on the positive side to see what you gain when you do them. Less stress, more free time, feeling good, or whatever – make it personal to you. When making your list, plan time for things that you really like to do. Fun should be part of every day and you should give it a high priority. Even if exams or emergencies take over your time, you can still plan things to look forward to after they have finished. It makes doing your lists more fun as well.

The easiest way to keep lists is always to carry a pocket notebook so you can keep track of things you need to remember. Some come with sleeves where you can keep things like your ID or bankcard too. That helps you remember to take it everywhere.

If your mates ask what you're doing, just tell them you're using an organising system. They may want to take it up themselves – we all forget to do things. Don't let them discourage you, because you really need to do this. They may even tease you about it, but that will stop after a short time. Laugh with them until it does.

What will you gain? Sometimes there may be a reward, like achieving a good grade, being thanked or congratulated, or earning money; other times people might not even notice. But to know you have done something well is the biggest reward. When you have completed all the tasks on your list to the best of your ability, write "well done" at the end. Congratulate yourself. Sometimes you may even give yourself a little present, or do something you like to do.

Don't leave any task unfinished unless you simply run out of time. If that happens, add it to tomorrow's list if it's urgent or to a future day if it's not quite so important just now.

Here is a checklist to help you learn to stay focused and organised and for improving your memory as will be discussed in the next chapter. Start to practise **"Think, Say, Do"**, a pattern that we will use with every important step. These steps will help you build habits to stay organised in spite of your ADHD.

1. **THINK**. Use your imagination to create a clear picture in your mind. **Think positive thoughts!**

2. **SAY**. Talk about it, write down, or draw it on paper. If you tell someone that this is where you are going to keep your keys, for example, you're much more likely to keep them there than if it was just an idea in your head. Writing it down stops you forgetting.

3. **DO.** Nobody gets good at anything without practise. Do things enough and they become a habit and you won't need to worry about them anymore.

Think about what you are accomplishing. Will you do it differently next time? Move towards your goal, trying out new ideas. Keep on until you succeed. Never give up. Remember that you are very creative. If you really don't know what to do or how to do it, ask for help. There are mates, parents, teachers, doctors and counsellors who can help you.

Focus, Remember and Organise
Think, Say and Do

❖ When you go for a walk make a note of some of the things you see, hear and smell. Check how many you can remember later. Make this a habit. Soon you'll start to notice more and more.

❖ Link people's names and faces with others you already know, or with things that come to your mind when you look at them.

❖ Think of funny and strange pictures for stuff you have to remember, like the sun shining on your textbook, a key stuck in a shoe, or a pen curling itself around a tree.

❖ Attach numbers to objects, like "one" and "sun", or "two" and "shoe". So, for instance if you have to meet someone at 2pm, you'll be reminded every time you see a shoe.

❖ When you need to concentrate, take deep breaths (not hyperventilating!), and relax the muscles you aren't actively using at that point – you will be a lot more focused.

❖ Put your keys, pocket diary, etc, in the same place every time. Never be tempted to leave them anywhere else.

❖ Use your mobile phone for reminders.

❖ Make set times for doing particular things and stick to them.

Improving your memory

Remember to remember and forget to forget!

We all have short-term and long-term memory. After you see or hear something, you remember most of it for a short while – for instance, when someone tells you their telephone number and you dial it immediately. If, however, something distracts your attention before you dial it, you may not remember the full number, and will have to ask again if you didn't write it down. This is a simple example of the very short-term memory failure, which affects most people.

Usually you'll remember quite a bit of what happens and what you see or hear. Your memory will even improve slightly within the first five minutes, but then it slowly starts to wane... although you will retain some of it, depending on how vivid your impression was. For instance, if you see an accident occur, it is often easy to recall the memory.

Learn to 'hook' to remember

Usually you have to repeat what you learn at school, or at least think about it a few times before it is stored in your long-term memory. That is why everyone needs to revise. It's better not to wait too long before doing revision; each of us needs to figure out how often we need to do revision so we don't have to keep learning the same thing all over again! Make your own summaries, and look at them regularly until they are stored in your long-term memory. Like everyone, you may need to look at them again occasionally to retain the knowledge.

It is much easier to retain knowledge if you can "hook" it onto something. A good example of a hook is to write different ideas in different colours. You will remember them better. For instance, when you summarise a historical event, write the main events down in blue, and the dates behind it in red. You must also have a clear picture of what you want to learn if you want to remember it better. Another very important hook when you learn new information is to associate it with something you already know. For instance, if you are introduced to someone called Ashley, you may remember someone with the same name and connect the two. It also helps to make mental pictures of things you want to remember. The more outrageous your mental picture is, the better you will remember. This also makes remembering fun, besides being practical.

Here's one I might have used to remember Surrey before I lived there. I picture a big man with a moustache jumping up and down saying "sorry" to someone he knocked over. Sorry is similar enough to Surrey that I could easily remember it. Of course, when you stay in Surrey, you will see and hear the name all the time and you will remember it without any effort. But if you lived in another county, you would need to have ways to remember it.

An interesting fact about numbers is that we are able to remember seven numbers or items plus or minus two (so also nine or five numbers). That is why telephone numbers are usually no

Sevens are easier to remember

longer than seven digits (without the code or the mobile prefix, which may be seen as another number). The fact that you are able to recall seven things comes in handy if you want to remember different items. This is because you are able to attach up to seven more items to each of the original seven. For example, if your diary lists football, you may need your kit to be washed (if your mum hasn't done it already), you may need new boots and so on, up to seven things to be done before the match. Then there're may be anything from one to six other items with their "attachments" to remember. It is much easier to remember up to 7 x 7 items rather than 49 items in one list.

You can find many more memory tricks like these in books or on the Internet, for instance, *"Mind Maps"* by Tony Buzan. Don't try too many methods, though; stick to one or two, otherwise it will become confusing.

If you have ADHD, you are like someone swimming in a river infested with crocodiles. You swim as fast as you can, but the crocodiles will catch up with you, because they are used to the water and are able to swim very fast. You might cope by scaring them away, for instance, if you shot at them with a gun, but you'd have to be very strong and fit to be able to fight them off on your own.

Being unable to concentrate or sit still, not finishing your schoolwork or homework, calling out the answer before the teacher has even completed the question, and forgetting or losing things (like keeping track of the finished homework until it's time to turn it in) are all like crocodiles that are after you when you have ADHD. You try very hard to fight them off or swim away, but they catch up. This is when "bullets" in the form of medication are needed – to scare them away for a while, so you can swim and make progress to the other side of the river. This translates to finishing school work or homework without being stressed or feeling harassed.

If your doctor has prescribed medication, remember to take it. It will make you feel more balanced and enable you to sit still and concentrate better. You may also have more peace, and become less angry. You have to take just the right amount at the right time. Talk to your doctor to help you with the schedule you need. Your doctor will also tell you and your parents of possible side effects from the medication. Report any side effects that really bother you – your doctor can change the dose or try another medication less likely to cause those side effects.

There is one more thing to remember about the crocodile story. Even if there are no crocodiles in sight, you still have to swim across the river or you will drown. This means you still have the responsibility to do your part. ADHD must never be an excuse for you to do nothing, or to be nasty or do nasty things to others. You have to take responsibility for your own actions, though others can help you. You also have to take responsibility for taking your medicine to help you feel good and get on top of things.

HELP FOR TEENAGERS

Getting Your Life Organised

ADHD often makes staying organised hard. It's hard for almost everyone at times, so it might be even harder for you. It is very important that you make and follow a diary.

Plan **a time for everything** – and **do everything on time.** To stick with that, you also have to have a **place for everything** and **keep everything in its place.**

You have to do things at the right time for your "systems" to run smoothly. If you don't understand this and do it, you will never feel in control of your life.

You need these systems to help you save time and keep track of things. They run by themselves if you set them up and follow them.

If you don't dedicate the time to do something, it usually won't be done, or at least not as regularly as it should be. Some things you have to do, and others you have to make sure that other people do.

For instance, studying regularly rather than studying for a test at the last minute is very important for you – you mustn't waste time rummaging for things when you need to do something, either; they must be in their place. Keep

Find a place for everything

your notes, homework and completed tests for each class organised so you can easily find what you need.

For things to run smoothly, you also need to turn up for activities like music lessons, sports practice, play rehearsals, or for a job if you take one. Keep track of things you need to do at home as well as at school. All of them need to be on your list or schedule, plus time for fun – things you like to do, hanging out with mates, etc.

If you have to decide between two things to do, look at each and ask yourself, "Is this thing going to help my systems run smoothly or is it going to mess up my plans?"

Just to go over this a bit more than we did in the introduction, here's the best way to set and keep schedules and stay organised and focused on what you need to be doing.

You need to have a diary somewhere, in which you will list plans, deadlines and schedules. This needs to be a proper diary, ideally divided into hours or half-hours. Use a pencil so changes can be erased. Booklet calendars that have the times of day and a spot for notes or goals on each page work best, and one that fits in a pocket makes it easy to carry.

Note your **deadlines** in your diary. **Don't allow deadline projects to pass by without being done,** even if you have to give up one of your other activities to make time. Better still; try to carry deadline projects out in advance. Don't go from one crisis to another. Ask for help if you need it.

Take a few minutes at the end of each day to plan for tomorrow (or at the beginning of the day if you are a morning person). When planning your day, set priorities for tasks. You may want to use 1, 2, 3, etc, to note things you must do.

Don't take on more than you have time for. Learn to say "no" when necessary. Try not to get so involved with one task that you allow your deadlines to pass undone. You need to keep a balance.

You may also take some time to **plan at least the week,** maybe the month, and even the year if you have an important long-term project or goal you want to achieve. This helps your life to have a steady momentum in a worthwhile direction.

Get Your Priorities Right

Decide which things are important and give more attention to them. This way, you will have more time to do these things, because you will not be wasting it on unimportant stuff.

There is a checklist on the next page. There may be some low priority stuff that could be done quickly if you have a bit of extra time, but never spend significant amounts of time on these if there are still things with higher priorities to do or finish.

Do things in small steps
If you find that there are things that should be done but never get beyond the bottom of your list of priorities, break these down into smaller chunks. Prioritise one or two chunks per day. These can also be "spare time" tasks for when you find you have a few extra minutes. If you know what you want to do, you avoid wasting these minutes doing things like flicking through telly channels, playing with your mobile, or staring at the computer.

Take time to relax... but just sitting around doing nothing takes up the precious time you've planned for both serious and fun priorities. If you don't stick with your priority list, you can easily get way behind just because you don't use your time well.

You may find you also want to use your diary to budget for expenses – many people with ADHD are impulsive spenders. If you always spend your allowance or any money you earn within a few days and stay broke for most of the week, this includes you.

Remember, for each item on the checklist:

1. **Think**, imagine, and see a clear picture.

2. **Say**, discuss and write down, draw pictures.

3. **Do,** practise and evaluate.

Before you carry out a task, come to a definite decision to do it, and picture yourself accomplishing it.

| **Planning and Priority Checklist** |
| **Think, Say, Do** |
| ❖ Make a place for everything (get rid of stuff you really don't need). |
| ❖ Put everything in its place. |
| ❖ Use a diary or reminders on your mobile phone, which you always carry with you. |
| ❖ Have a calendar for planning in your room and mark important dates on it. |
| ❖ Schedule a time to make your lists and plans. |
| ❖ Schedule time for "must do" activities. |
| ❖ Schedule time for leisure activities. |
| ❖ Prioritise activities, using 1, 2, 3 etc. |
| ❖ List other things to be done at some time. Break them down into chunks and add them a bit at time to the priorities. |
| ❖ Plan – by the day and week, longer for really big, important stuff. |

HELP FOR TEENAGERS

Improving your concentration

Mate, lets concentrate!

With ADHD, your ability to concentrate on anything for a length of time is likely to be bad. This, plus the intensity of your concentration (varying according to how interested you are in that particular thing or topic) can cause difficulty with memory. You will also tend to be more likely than the average person to forget what you were doing or thinking just before a distraction. That's usually a minor problem, but not always.

New drivers

Scientific studies have found that all new drivers are much more likely to have accidents than people who have driven for at least a year or two. They recommend that new drivers start out without any passengers (unless, of course, the passenger is teaching you to drive) and then build up slowly, getting used to one passenger, then two passengers, etc. Some recommend that you do not drive with more than one passenger during the first year you become a driver. Knowing that you're easily distracted, this is an especially good idea for you.

When driving, you really have to concentrate Switching your attention just briefly can be very dangerous if you are riding a bike or driving a car – see 'Driving with ADHD' on page 88. The same is true if you are using power tools or working with any moving objects.

There are things you can do to keep your attention on one task. "Looping back" your attention means making an effort to stay in contact with your initial activity. If your mind wanders off to something else, it shouldn't become a separate "circle", but instead a "loop" back to your first thought. This habit, which you can teach yourself, involves going back to the original point where you were distracted.

We've all started out to do something and been distracted by something else before we start, and then forgotten what we originally planned to do. Sometimes if you go back to where you started, you will remember what you wanted to do. Use this in other situations. Go back in your mind to the moment when you decided to do something and picture what you were doing. Your thought will often come back.

Try to concentrate on one thing at a time. Keep in mind what you are going to do next if it's related to what you're doing now, so you don't interrupt the flow of concentration. If your next task doesn't relate to what you're doing now, jot it into your diary (if it's not already there), so it doesn't distract you from your current activity. Always have another activity "up your sleeve" if you get stuck with the first one for a while, so you don't waste time while waiting.

You may also have "internal distractions" over which you have less control. These may be due to tiredness or emotional issues. For people with "Inattentive" ADHD or ADD, attention deficit may frequently be internal as well as external. Daydreaming may be an important component of internal distractions. Of course, you don't want to daydream in class or at work, but to have dreams about the future is a good thing. *"Dream, but don't let dreams become your master"* is a famous thought paraphrased from a poem by Rudyard Kipling.

A word of caution: If you find that, you "switch off" at times and lose contact with your environment, or frequently get an overpowering need to go to sleep, talk to your doctor about it so he or she can help.

Try not to carry home issues to school or work, and vice versa.

Take regular exercise, try to get enough sleep, and stick to a healthy diet. Give yourself "a reward" for a task well done. Make it fit the size of the task. It could range from, say just getting up and walking around for a while, to a night out with mates, or doing something you especially like but can't always afford or find time for. Rewards will help you finish your tasks within the time you've allowed for them and help to keep distractions away.

Texting, emailing, Facebook, Twitter and YouTube videos can be major distractions that ruin your concentration and eat up your time. There's no reason why you should miss out on these things but, just like everything else, you need to schedule a time for them or important things will never get done.

If you have to do something that doesn't interest you, one technique is to turn it into some sort of game. For example, if maths is hard, see if you can think of it as a puzzle you will enjoy having solved. If reading is hard, be sure to read some things you enjoy. This will improve your skills and make things you have to do go faster.

Remember, for each item on the checklist:

1. **Think**, imagine, and see a clear picture.

2. **Say**, discuss, write down and draw pictures.

3. **Do,** practise and continually evaluate (Think about what you're doing).

Improving your concentration
Checklist
❖ Get everything ready before you start anything.
❖ Keep what you need for the current task to hand.
❖ Follow a specific programme and a steady flow of activities.
❖ Eliminate external and internal distractions as much as possible.
❖ Concentrate on one task at a time, but keep the next one in mind.
❖ Always "keep another task up your sleeve" to carry on with if you get temporarily stuck on the first one.
❖ Give yourself rewards for completed tasks, and don't try to work all the time.
❖ Follow a healthy exercise, eating and sleeping plan.

HELP FOR TEENAGERS

Controlling negative emotions

This chapter shows you how to enjoy life effectively and allow your emotions to function within their limits, the importance of your own self-image, and how to manage feelings of inferiority, frustration and depression. It also includes some anger management strategies.

We all have a self-image that is with us every moment. How we think about ourselves determines to a large extent our ability to relate to others. You can't be yourself if you're worrying about what someone else may think about you all the time.

There are many things that may affect our self-image. For instance, being continually told off by others may make us shrink into ourselves and feel rejected.

Concerns about what others may think of us can lead to worry and depression. We may allow our self-image to be shaped by the opinions of others or, more often, what we believe other people think about us. This makes us very sensitive to criticism and affects our personality, which is largely based on our self-image. Then, once we start living in the shadow of the image we think others cast on us, we can literally start to become afraid of ourselves.

Teenagers are especially concerned about what others think about them. The fact is that other people don't pay nearly as much attention to you as you might think. They're too busy worrying about what you think about them, so try to relax and just be yourself.

Another attack on your ability to have a healthy self-image is the feeling of failure, which can be pretty common with ADHD. When you find it difficult to concentrate and apply yourself to a task at hand, it's not easy to do your best, and you may get the feeling you are not good enough. This is especially true if you get into trouble for delivering poor work, which you know you could have done better if you had more concentration or more time.

Once a negative self-image gets hold of you, it starts to grow stronger by projecting itself onto other people. You start to believe that it's the real you. Very often it is the worst you that you could imagine yourself to be. Even though in reality, it's just an illusion that you have conjured up in your mind, fighting back to realise all the positive qualities you have is never easy. A lot of the exercises in this book will help you take back control of your life.

Taking the first step is often the hardest, but most important is for you to believe it's possible.

Here are some guidelines to get you started:

I have already mentioned this, but it's so important that I say it often. Say to yourself:

"I am going to be the best me I can be."

Then follow with some more positive statements like:

"I am going to use everything that has been given to me – my ability to think, my eyes, my ears, my hands and feet, and my personality, which is unique."

"I am not going to compare myself to others, or allow them to compare me with themselves, because I only compare myself to the best I can be."

"I will try not to make mistakes, but I will not be afraid to make them. I'll use them as stepping-stones to do better next time and the time after that."

And, very important:

"I am not going to stress!"

It may not be easy, as stress "comes by itself," but it can also "come by habit", which you can break by positive thoughts and actions. Stress drains your "brain energy" and dulls your senses. Doing things which have to be done ASAP also relieves stress and anxiety. Procrastination and stress go hand in hand.

Did your mum teach you this rhyme when you were younger?

"Let the sun shine in, Face it with a grin,
Smilers never lose, Frowners never win..."

(Stuart Hamblen)

HELP FOR TEENAGERS

Relationships

Relationships are possibly the number one reason for **happiness** and **unhappiness.**

Relationship problems are usually caused by:

- ❖ Misunderstanding the feelings and intentions of others.

- ❖ Being dishonest or inconsistent.

- ❖ Doing things without sharing them with those who ought to know (like changing your plans about where to go without telling a friend you arranged to meet).

- ❖ Not doing what you said you would do, breaking promises.

There is no quick fix for these issues. You just have to learn from experience and observations, and on the way you will make mistakes. Everyone sometimes makes mistakes in relationships. Never see yourself as a failure, but instead learn from it. Here are a few suggestions to help you.

* **Listen to people**. When you are with someone, give him or her your full attention. Look at their face and body language, and listen to what they are saying – the words and the meaning behind the words. Listening is a difficult skill for a lot of people, but you will gain experience and get better at it if you try. Once you really start to listen, you'll see how much other people really appreciate it.

* **Say what you mean.** Don't expect others to know what you want to say if you don't say it. They aren't mind readers! It's easy to feel hurt and rejected because you think other people don't care about you**. But most often they simply don't know what you want or how you feel if you don't talk about it.**

* **Be honest.** Don't say what you don't mean. But don't say things that you know will hurt or annoy – even if they are true – unless if you have a good reason.

* If something that will hurt or annoy another person must be said, try not to say it in front of mates, or anyone else. **You need friends, not enemies.**

* If you want to praise someone, do it with honesty. If you just flatter someone to win their attention, they'll soon see through it.

* Try not to talk about someone when they are not in the conversation, because that can so easily be perceived as, or turn into, gossip. If the conversation turns in this direction, don't participate, and walk away. It's very easy to be drawn into gossip.

Dear Dr Marius,

As I am now turning 18 and leaving the 'paediatric age group,' I would like to write you this letter of sincere thankfulness for the years you have 'looked after me' after I was diagnosed with ADHD when I was 12 years old. Your understanding and talking to me about my problems and how to overcome them will always remain with me. I particularly appreciate the fact that even at a younger age you always took time to talk to me personally and not only to my parents as some doctors do.

John

Sometimes it's very difficult to tell the truth, especially if you have been told off many times in the past. But it is always better to tell the truth, and take the flack, rather than to lie about something. The truth has a funny way of coming out in the end, and any lies you told will be embarrassing.

Integrity goes with honesty. This means to be true, and true to yourself, at all times. It also means that you take responsibility for your own actions, and do not blame others for them. If you hit someone's car in an accident, or even just scratch it slightly, take responsibility. Don't look around to check whether or not someone saw you, and drive away if not. That's breaking the law.

The same factors apply to relationships. If you "scratch" someone, apologise and try to make amends for any hurt feelings or actual damage you've caused. If the other person is still upset, then let it remain their problem – not yours. The most important step you can take on your road to being a person with integrity is to keep your word (promises) every time. Even if it hurts right now (such as wanting to break a commitment to do something with a mate because something else you'd really like to do comes along), in the end keeping your promise will repay you in having good relationships.

Dealing with anger

❖ Do you get angry easily and often? As a child, you may have been told off and rejected frequently. This made you feel angry, and that anger may have become a habit. This is a habit you want to break – it takes a lot of your energy, and can push other people away from you.

❖ Don't expect more from others than you expect from yourself. Keep only "short accounts" of what others do to you. When you forgive someone, it brings even more healing to you than to the other person. If you have said or done something that affects a relationship, saying that you are truly sorry about it usually works, even if it takes some time for the other person to forgive you entirely. After you apologise, don't carry guilt around – just let it go.

❖ Try to stay calm when things go wrong. Don't lash out and try to protect yourself by attacking others. Rather, see how the problem can be solved, and do the best you can in the circumstances. This is not to say you should let others "walk all over you." Stick to your principles, but try not to over react.

❖ Don't qualify an apology by telling the other person what he or she did. If there are issues that you need to let go of, deal with these first before apologising.

Personal space

Another problem people with ADHD tend to have is intruding on the personal space of others. Be careful not to come too near or be too familiar with people you don't know very well. These are things everyone needs to learn, but people with ADHD need to pay special attention to them, because you may have a little more trouble reading emotions than many other people.

Here are a few tips for keeping relationships healthy and positive:

❖ It will help you stay balanced if you try to see the good in others and respect them for that, even when it's not easy.

❖ If you are in a position to help someone, do it just because it feels good, not because you expect to be thanked.

❖ Try to surround yourself with positive people who will pull you up instead of pulling you down. Sometimes this will mean that you actively have to seek the right friends, and unfortunately, also learn that it's smart to spend less time with others.

Remember, for each item on the checklist:

1. **Think**, imagine, and see a clear picture.

2. **Say**, discuss and write down, draw pictures.

3. **Do**, practise and think about what you are doing.

Checklist for Good Relationships
Think, Say, Do

❖ Understand that communication is important and try to be genuinely interested in others.

❖ Find people who like some of the same things that you do.

❖ Do things together with others, even if you think some activities may be stupid or boring – there is no better way to build relationships.

❖ Have your own view, but allow others to have theirs as well.

❖ Learn "body language" and work on reading it.

❖ Do what you say you will do, and keep your promises.

❖ Try to be honest and consistent with other people.

❖ Admit when you are wrong and apologise, but don't go around loaded with guilt.

❖ Look after yourself well (be clean and neat).

❖ Deal with anger. You may need help with this. Talk to your doctor or a counsellor if you need help.

HELP FOR TEENAGERS

Learning, coordination and perception

Because you have ADHD, the chances are you've found out that some of the ways they teach students to learn and remember at school may not work well for you. If so, by now you are likely to have worked with teachers and counsellors to find how you learn best. As you become more independent in school and everywhere else, you need to take on more responsibility for learning. You may have been able to count on your teachers to adjust things for you when you were younger, but now you need to become responsible for applying the ways you learn best. Even if you have already "learned to learn" successfully on your own, you may hit a bump and feel as though you are not learning or can't follow what is being taught. Ask for help from your teacher, counsellor or doctor. When you change schools, it's a good idea to let your new teachers know about your different style of learning so they will be ready to help if needed. Your parents can help with all this too.

Sometimes you need to learn to learn

ADHD may be associated with specific learning difficulties, meaning you can probably learn well in many subjects, but may have trouble with specific subjects or skills. Your teacher or a special needs coordinator can help with this.

Some people with symptoms of ADHD have problems with motor coordination. This can be hard at school, and have a negative effect on your self-image. Or you may be good at some sports, but not others – just like everyone else. You may be left out of some teams, which may make you feel angry.

Find people who are interested in things you do well and do things with them – maybe you sing, play an instrument or are good at art or science, for example.

Laugh at yourself

When you grow up you can choose your own sports and activities. You may continue to be clumsy at some things. Try to avoid difficult situations whenever possible, and learn to laugh at yourself. You'll feel better because laughing increases those positive substances within you that are often low when you have ADHD.

Poor perception, either hearing or seeing, may be a problem. If you have been tested and your ears and eyes are fine, you need to set about finding ways around it, such as looking and listening more intently, or avoiding activities which you know you won't be good at. Teachers and occupational or speech therapists are the best experts for this kind of help. Also be aware that almost everyone has some problems they need to work around. For instance, imagine looking at a bird in a tree if you were colour blind. About one in every 76 people has some problem with colour blindness.

If you have started driving and frequently get lost, you may need to ask your parents to invest in a satellite navigation system for you. You might also find and follow a good map, or have someone travel with you to check directions as you go. Fortunately, like anything else, once you have followed a route several times it will become easier to remember.

You may find yourself growing out of some symptoms associated with ADHD as you go from being a child to being an adult. With the skills you learn now, you'll be ready to cope with the symptoms you keep and things that come along in life.

Keeping going

What are you good at? Start spending more time on something you like doing, and improve it. Tell yourself:

"I will use my brain, my ears, my eyes, my hands and feet to be the best me I can be. There is no one else on earth exactly like me."

"I will be the happiest me I can be."

Put one foot in front of the other and if you look back after some time you will be amazed at how far you have travelled. It's never too late to start. Keep on going. It's not how something begins that counts, but how it ends.

If you need to decide something, don't put it off. Get as much information and advice as possible, and then make your decision. Move away from a "crossroad" that creates anxiety. Act on and trust in your decisions.

When you make a mistake, see if there is anything you can learn from it – and then do it differently next time.

Ask yourself, "How long am I going to feel bad about it?"

Then answer yourself with, "Maybe 10 minutes, or half an hour!"

For this time, feel really bad; then tell yourself, "Stop!"

There is a chapter on "Developing and maintaining momentum" in the adult section of this book which you may want to read. One "experiment", which I mention there as well, shows how you can build up momentum through better organising things. Next time the kitchen seems to be in a mess, tell your mum you will clean it up or you will help if you can be in charge.

First clear and put away everything that does not need to be on the worktops or in the kitchen, including leftover food that can be discarded or stored for the next meal.

Then tackle the dishwasher if you have one, unpacking and packing it in a certain order; e.g. bigger things first, plates and bowls, glasses and cups, knives, forks and spoons. Then wash big things that don't fit in the machine, and do a final clean up, including taking out the rubbish and sweeping, vacuuming or mopping the floor if necessary.

Each thing you do takes away a visible chunk of the job, which then looks less daunting. To make things more interesting you could time yourself and see how much sooner the kitchen becomes tidy when you use this method. You can apply this system to any task, even to your homework! **Remove visible chunks again and again until nothing is left and see what has become of the "mess" when you look back at the end.**

Motivation and goal setting are key factors in keeping momentum. Also: Discipline, Commitment, Attitude, Fitness (both physically and mentally), Integrity and the Dedication of time for important tasks or projects.

Here is a simple mathematical formula to help you:

KNOWLEDGE (PLANNING) (5) + EFFORT (5) + TIME NEEDED (5) + ATTITUDE (RELATIONSHIPS) (5) + CONSISTANCY (5) = RESULT (25)

You can use this formula to evaluate the outcome of any task by giving points for the result. Then look at the items on the left side of the equation, and decide on which items you are losing points. Try to improve on these.

Momentum – keeping going – is dependent on discipline rather than feelings, though feelings may affect it. The speed that you walk, for instance, may be affected by your mood but, even so, if you keep putting one foot in front of the other, you will still cover the distance. This is not to say that you must live your life by some rigid framework; a healthy self-discipline allows for some flexibility – but not an excessive amount.

Checklist to Keep Going
Think, Say, Do

❖ If something goes wrong (especially in relationships), correct it as best you can, limit time to feel bad, and then stop!

❖ Wherever possible, mix with positive people who build others up, and don't tear them down or gossip.

❖ Make the best of every situation.

❖ Make time to do what you enjoy and can do best.

❖ Come to terms with things you are not able to do.

❖ Tackle overwhelming jobs 'one chunk at a time' so you can see them getting smaller.

❖ Cultivate a healthy self-discipline so you are not ruled by your feelings.

Remember, for each item on the checklist:

1. **Think**, imagine, and see a clear picture.

2. **Say**, discuss and write down and draw pictures

3. **Do**, practise and continually evaluate.

Specific treatments

There are lifestyle and medical treatments for ADHD. You may want to try just the lifestyle changes, maybe including counselling, but medical treatment and medications may also be advised. Medical treatment along with lifestyle management, and strategies to overcome some of the behaviour problems – remember the crocodiles! – can make your life much happier and more relaxed.

You may find that you are outgrowing your ADHD, though it can be hard to figure out during your teenage years, which are a roller coaster for you and everyone else, as your body and brain complete their growth and you move towards becoming an independent adult. If you don't outgrow your ADHD, this book can still help you to become – and stay – a balanced and successful person.

Since you have ADHD, you must be wise and work on the problems it can throw in your path. You need to learn to control anger, and may be unusually defiant to your parents and teachers (which is as hard on you as it is on them) and do things against your better judgment just because you think they won't like it. If you don't work on treating ADHD, your teenage years can make it worse instead of better. You're much more likely to underachieve at school, and to get into various kinds of trouble. Teens untreated for ADHD average two arrests by the time they turn 18. About 20% of teens untreated for ADHD will eventually be arrested for a felony, versus only about 3% of teens without ADHD. You must be aware of the facts that you are more likely to become depressed, resort to alcohol or drug abuse, and you need to let your parents and/or doctor or other trusted adult know about any of these kinds of problems before they become severe. See "Talk to Frank" under "Useful Resources" at the end of the book.

Your doctor can prescribe medications that help keep you balanced and less restless and distracted. Usually they are "stimulants" like methylphenidate in its different forms (Ritalin is one), and dexamphetamine. There is also a non-stimulant medication called atomoxetine (best known under the brand name of Strattera in the UK and Paxil in the US). I am not going to burden you further with details, but enough to say they increase the neurotransmitters, which we talked about the section on "Understanding ADHD". Your doctor will explain what a medication should do, how it works, how often to take it and possible side effects when prescribing it.

Make sure you say your "best me" sentence to yourself at least once a day and remember it in any difficult situation:

"I will be the best me I can be!"

And remember to have some fun every day.

If you find those ADHD crocodiles chasing you, be sure to review this book, and then talk to someone if that's not enough – doctors, parents, counsellors, teachers and mates can help.

THE TRANSITION YEARS
ADHD changes from teen to adult

In general, teenagers want the privileges but not the responsibilities of the "boring" adult world. However, just as the world is turning, so the times and circumstances (from school to college to work) are changing, and the responsibilities of growing up just become more "intrusive". As the "transition" to the adult world steadily becomes a reality, teenage interests and attachments become weaker and more mature ones replace them. This happens gradually and some people may even get "stuck" at a particular stage.

For someone with ADHD, the "environment" changes rather than the condition. Hyperactivity in the child is replaced by restlessness but inattention and impulsivity stay very much the same.

The interaction between the person and his or her environment is the crucial, underlying, and ongoing "cause" for normal and problematic development, with or without ADHD.

Impulsiveness	
CHILD AND TEENAGER	ADULT
1. Often blurts out answers before questions have been finished.	Does not wait for the other person to finish a question before giving an answer. The question is completed in his or her mind, but may not be what the other person was asking.
2. Often has trouble waiting his or her turn.	Difficulty following rules; wants to take "short cuts" all the time. Gets very irritated when having to wait; "as if the only person in the world."
3. Often interrupts or intrudes on others (example: butts into conversations or games).	Interrupts or intrudes on others while they are still talking or busy doing something. Seems as if he or she just wants to express their opinion without even listening to what someone else may be talking about.

Notes to the tables

These tables are based on the 18 symptoms of the DSM IV criteria for the diagnosis of ADHD in children. (These criteria should only be used by professionals with the necessary expertise to diagnose ADHD and not for self diagnosis.)

The DSM IV classification does not have specific criteria for the diagnosis of adult ADHD (due to be incorporated in DSM V). The adult symptoms in the table are a compilation from different sources, including my own experience.

The (+) gives an indication of how you could adjust positively (either by yourself or with the help of someone else) to an important aspect of the issue. There are just three examples given. Read the teenage and/or adult sections of the book to learn how to adjust any of your other symptoms.

Hyperactivity	
CHILD AND TEENAGER	ADULT
1. Often fidgets with hands or feet or squirms in seat.	Can't sit still without moving something e.g. head or shoulders, wringing or rubbing hands, reclining and then sitting up straight; constantly moving legs. (This is called "restless leg syndrome, which is usually hereditary and may continue at night causing disturbed sleep both to the sufferer and their partner as well).
2. Often gets up from seat when remaining in seat is expected.	Has difficulty staying seated for longer periods; may find sitting through sermons, lectures and plane journeys distressing, or even like "being tortured."
3. Often runs about or climbs when and where it is not appropriate (adolescents may feel very restless).	Feels restless and may become irritable easily. This, together with impulsiveness, may lead to risk taking acts and "self-medicating" with nicotine, alcohol, other drugs, pornography and irresponsible sex.
4. Often has trouble playing or enjoying leisure quietly.	Doesn't like work or games that require sitting quietly.
5. Is often "on the go" or often acts as if "driven by a motor".	Always wants to do things; always on the go. Can't stay at home.
6. Often talks excessively.	Feels a constant need to talk, and can't stop talking. Doesn't give others a chance to talk. This may lead to saying inappropriate things that unintentionally offend others.

Inattentiveness	
CHILD AND TEENAGER	ADULT
1. Often does not give close attention to details or makes careless mistakes in schoolwork, work, or other activities.	Doesn't want to "waste time" on detail. Would rather follow than gather enough information to lead; tends to make "uninformed decisions" and not "read the small print" (though sometimes gets into the details and forgets what's important). Makes wrong judgments regarding financial issues, which may lead to being easily exploited. (+) Make *informed* choices/decisions. Small details may "grow" big as time goes on.
2. Often has trouble keeping attention on tasks or play activities.	Has difficulty keeping focused on the task at hand. Thinks about non-related things. Loses track in lectures or meetings and may "doze off" or day-dream frequently. (+) Adhere to a sleep and exercise program. May need medication.
3. Often does not seem to listen when spoken to directly.	Seems not to give attention to what others say, to their annoyance, which can lead to problems with relationships, especially marital. May be accused of "not caring." Has difficulty following instructions. (+) Get more interested in the other person's interests/needs. Ask someone who understands to explain your problem to the person, but don't take advantage of it.
4. Fails to follow instructions and finish schoolwork and chores (other than ODD or failure to understand).	Becomes bored with a task after only a few minutes, unless doing something that is enjoyable or exciting; starts projects enthusiastically, but does not finish them; lacks ongoing motivation.

Inattentiveness

CHILD AND TEENAGER	ADULT
5. Often has trouble organising activities.	Is disorganised; has to look for things all the time; misses deadlines; is usually late for meetings; is reactive instead of proactive; fails to plan ahead; functions by "crisis intervention"; has a nagging feeling that tasks are never well done or complete.
6. Often avoids, dislikes, or doesn't want to do things that take a lot of mental effort for a long period of time (such as school work or homework).	Gets "tired" of doing mental work for any length of time; procrastinates; avoids and dislikes tasks or activities where you have to "think" a lot, such as research or reading non-fiction. May be perceived as "lazy." Doesn't want to bother to be selective in what has to be done.
7. Often loses things needed for tasks and activities (such as toys, school assignments, pencils, books, or tools).	Is inclined to lose or misplace keys, important documents, mobile phones, messages and telephone numbers written on pieces of paper. Has a poor sense of direction; gets lost easily.
8. Is often easily distracted.	Is easily distracted by internal and external stimuli. (Together with impulsivity this may be the reason for a higher incidence of accidents.) May switch attention when distracted and forget to "switch back."
9. Is often forgetful in daily activities.	Forgets important dates and commitments, and where things are put. Can't remember names. Is sometimes accused of "using forgetting as an excuse for not wanting" to do something.

HELP FOR ADULTS
Your executive functions

"If you work hard at your work, you will become wealthy. But if you work hard on yourself, you will become a millionaire!" (Jim Rohn)

Life is no fun when we never pay our bills on time, start projects that never get finished, waste hours looking for things, and frequently upset the ones we love by forgetting important dates and to do the things we promised. If this sounds all too familiar to you, then the following chapters can help get you get organised and in control of your life.

When treating children and teenagers for ADHD, advising the parents on how to deal with their children has always played a very important role in the treatment program. However, it is quite common to find that one or more of the parents has problems with ADHD as well. About two-thirds of the adults who had ADHD as a child still suffer from the condition to some extent, and only half of these have learned ways to overcome it. The other half needs ongoing support and possibly medical treatment. Unfortunately very little assistance is presently available for adults with ADHD though there are some centres or individual psychiatrists who offer help.

The advice in this section of the book is based on both my experience as a doctor and on my personal experience as an adult with ADHD. It utilises motivational techniques that have been proven to work both for my adult patients and myself. The advice is not very different to that which I give teenagers but it is geared more to the adult world and all the extra responsibilities and expectations that come with it.

Your inner manager

We all have an inner manager that helps us to adjust and cope in different situations and environments. This is true for children, teens and adults, with or without ADHD. The role of the inner manager is to orchestrate what are termed the executive functions (indicated in the diagram opposite). Your working memory, your feelings of success and failure, and your abilities to organise, concentrate and remain attentive are all based on how well you can adjust and balance these executive functions to cope with different situations and environments.

The way I see the "inner manager" is that it manages and influences both the conscious and subconscious mind through various tools, such as positive emotional responses, for example, that trigger us to start and finish a project. When it doesn't function properly, you may be left with the nagging feeling from your subconscious that you ought to start or finish a project. Think of your subconscious as your "genius self", which remembers your past goals, dreams and aspirations for the future.

When your "inner manager" is impaired you are like someone with their hands tied behind their back unable to use their tools properly. In this case, your "chemical tool bag" is your working memory, feelings of success and failure, etc. Using "rational will power", you are able to modulate your behaviour on the conscious level, but this will have only a limited influence on the subconscious level as you proceed with an activity.

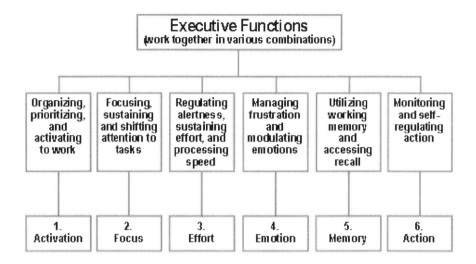

A description by Dr Brown of the different aspects of the Executive Function is presented in his book: "*Attention Deficit Disorder: The Unfocused Mind in Children and Adults*".

Without the "inner power" or feeling to accompany your rational will to do things, it's hard to keep motivating yourself and life becomes a struggle. You need put a disproportional amount of energy to keep yourself motivated and focus on the task at hand.

The solution to this problem is firstly to increase the levels of neurotransmitters in the areas of the brain that support the Executive Functions and secondly to develop certain patterns of behaviour to educate the brain to overcome these shortcomings.

Exercise or receiving other exciting visual or auditory stimuli – the playing of video games for instance – can raise the level of neurotransmitters in a person with ADHD. Even memories or fantasies about these experiences can induce elevated levels of these "internal stimulants." However, a sufficient level of blood or tissue concentration, no matter whether it's produced by the body or taken as a medication, only lasts for a while, and has to be produced or taken regularly to keep the level sufficient. As mentioned elsewhere; in the person with ADHD the "default level" of these neurotransmitters is lower than needed. It has to be continually replaced or stopped from being excessively re-absorbed for the person to function optimally.

The production of neurotransmitters (especially dopamine) in the body may have an addictive effect. If one type of stimulus is allowed to develop out of proportion it affects the balanced need of the person for other stimuli. For instance, if a person receives excessive stimulation (and produces high levels of dopamine) by watching pornography, the experience may become addictive, which can make it very difficult to switch to other interests and has an overall negative impact. This may be illustrated by the "trampoline effect". If you jump on a trampoline and then get off, your body feels as if the power of gravity has been increased many times, and immediately afterwards it is difficult for you to jump higher than a few inches from the ground. This shows how difficult it may become for someone who has been excessively stimulated by one activity to switch to another.

General excitement and enthusiasm about life (the adventure factor), a balanced physical exercise program, a healthy diet and enough sleep, will go a long way towards increasing the release of neurotransmitters. In some cases, however, this may still not be enough and you may need to get medication from the doctor. Don't see this as a crutch or an excuse to give up on the other things. Think of it as just a part of the toolkit that will keep your life in order.

Knowing your limits

As an adult with ADHD, you have to realise that you are not always the best person to manage certain areas of your life. For instance, if I had let my wife manage our finances, or at least been more open with her about them, we would have been in a much better financial position than we are now. Delegate or share the responsibility for the things you are not so good at, to a partner, a trusted friend or a professional service. Knowing your limitations is a sign of wisdom not weakness. Never be too proud to ask for help when you need it.

Don't be ruled by your emotions

It easy to say things like, "I just don't feel like it" or "I just don't feel good about it". Maybe you don't feel positive about doing something; you may even have a bad feeling from the past holding you back. But making a simple job like washing the dishes into a major chore isn't going to make you feel better. In the adult world, rational decisions have to made and acted on all the time regardless of feelings, or our society would grind to a halt. The good news is that once you begin to act, feelings generally take a back seat and can be kept there until the job is done. And, because you have accomplished something and didn't give in to your feelings, you are likely to feel much better. Feelings simply motivate us to do or not to do a task. Ultimately, it's you, not your inner manger, who is responsible. Have you ever noticed how fast a feeling can change once you take the lead? Part of maturing emotionally is being able to let your feelings know who is in charge.

Be the best me I can be

You may have noticed that I've adopted this as a personal motto and I recommend others to do the same. It should never be seen as a selfish ambition, but a goal for you to develop your full potential to be the best person you can possibly be.

HELP FOR ADULTS

Getting organised

There are two forces in the world: **chaos** or decay on one side, and **order** on the other side.

Anything left on its own without the application of order will eventually deteriorate. A piece of iron lying outside will rust, and anything organic will decompose.

Take a car for instance… If you just drive it without doing any maintenance, things will start to go wrong. Eventually it won't run properly and it will stand outside. After a time, it will just become a carcass of a car; in a few hundred years, nothing will be left of it.

Exactly the same principle applies to our minds and bodies. If we don't apply order to them, they start to deteriorate and stop functioning. Fortunately, you can overcome this by getting organised and developing a goal-oriented master plan and road map.

If you seriously adhere to the following, I can assure you that you will become more organised. It may sound simple, but it's true. It may sound like extra work, but it will save you hours and hours of "catching up" later and time wasted in looking for things. Most importantly, it will give you the feeling of being in control.

A place for everything and everything in its place

Even if you do nothing else, this on its own will bring a change for the better. First find or make a place for everything, and get rid of or store things for which there is no place. This is of critical importance. Then see to it that everything is kept in its place as often as possible.

If you walk from one room in the house to another, **never walk empty handed.** Always carry something that belongs in the other room, like a glass or a cup in at least one of your hands, and teach the others in your household to do the same. Even if you come from your car, bring something to put in the rubbish bin. If you are taking your hands from one place to another anyway, why not put something into them that isn't in its place? Make it a rule to put away everything you use and clean up afterwards. The idea is not to have this as a burden or obsession, but as an organised way of life.

Schedule time for everything and do everything on time

You have to do things at the right time for your "systems" to run smoothly. If you don't understand this and do it, you will never have the experience of being in control of your life. Systems don't run by themselves, they are dependent on you doing certain things on time. **If you don't dedicate the time to do something, the chances are that it won't be done,** or at least not as regularly as it should. Some things you have to do, others you have to make sure that other people do.

For instance, to keep your car on the road, you not only have to put petrol or diesel in and check the oil regularly, but you also have to take it for a service, get an MOT, and pay the insurance and tax on time. If you hear a knock in the engine, it's more likely to get worse rather than go away, unless you do something about it quickly.

You mustn't waste time looking for things when you need to do something. They must be in place. For instance, know where to find the telephone number of the plumber to call if the toilet system is blocked.

For things to run smoothly, you also need to turn up for appointments on time, pay bills on time, make phone calls on time, water the plants in the garden on time, pay taxes on time, plan your holiday on time, and so on.

If you have to decide between two things to do, ask: "Is this thing going to help my systems run smoothly or is it going to put a spanner in the works?"

Make notes of **deadlines** in your diary. For example, note down payments you have to make, and create a time for paying them in your daily schedule. **Don't allow them to pass by without being done,** even if you have to sacrifice one of your other activities. Better still; try to do important things, like paying bills, in advance. Don't go from tackling one crisis to another. Ask for help if you need it.

When deadlines pass, you may think they have gone away for some time, but they usually come back with a vengeance! The old saying, "a stitch in time saves nine" is still very true. I won't apologise for repeating myself on important points like this one.

Take a few minutes at the beginning of the day to **plan it** out. This will save you hours of hard work on less important things. You may also take some time to **plan the week, the month and the year.** Schedule some time to do your plans. If can't be done at home, go to work a little earlier and do it there.

For time management purposes, you will need a diary – either in book-form, or one on your computer or mobile phone – divided into hours or half-hours. Fill it out with activities, including periods of rest and recreation. Then you'll need to decide your priorities. You may have to "borrow" from times that you have allotted to less important things to be able to meet your deadlines. Don't take on more than you have time for. Learn to say "no" when necessary. Try not to get so involved with one task that you allow your deadlines to pass unmet. You need to keep a balance.

Be prepared for emergencies

1. Keep emergency numbers close by where they can be easily found, especially when you aren't at home. Include the telephone numbers of the **ambulance, fire brigade** and the **police**, along with you GP, local hospital and dentist.

2. Keep the number to dial in case of a breakdown of services, such as the **electricity and gas** (including the number to report any **gas leakage).** Know whom to call if there are problems with **electrical appliances** or with the **plumbing.**

3. Keep the number for a **breakdown service** for your car at hand, along with your **insurance company** details**.** Get a **spare key** made today so you aren't locked out.

4. Keep a back-up copy of all key numbers on your mobile.

Be selective and get your priorities right

Decide which things are important to you and give more attention to them. This way, you will have more time to do these things, because you will not be wasting time on unimportant stuff.

Develop a method for doing this. First write down all the tasks you **have to do,** and then mark them with their **level of priority**, (e.g. P1, P2, etc), or non-priorities (e.g.: NP1, NP2, etc). Delete the "N" from the "NP" if the task becomes more urgent.

Make sure you **always finish the priority jobs before working on the non-priority jobs.** There may of course be aspects of the non-priorities which can be done quickly (within a few minutes), or delegated to save time, but never spend significant amounts of time on NPs if there are uncompleted Ps.

Tasks that will be **nice to do or of possible importance** should be marked **L** (for like), and filed to be given attention later or moved up to **P** or **NP** status when appropriate.

Apply Pareto's 80/20 rule: do the 20% of activities that bring 80% of the results (or at least those that will bring you 80% of the grief if you don't do them).

There are certain jobs that will seem to keep you busy forever, and may take up all your time every day. You must take some time off from these responsibilities to do things that are more pressing and important. Once these are done, you can return to the ongoing tasks.

At home and at work, make a list of all the **outstanding things** that need to be done, or are waiting to be done, including those that have already been waiting a long time for completion. Break these tasks down into the smallest chunks possible. **Prioritise one or two items per day**, and complete them to get them off your list.

It's not very rewarding to start a project and then leave it for weeks or months. Even if you carry out just a small part on a regular basis, it will keep the flow going until it is completely finished. It may be possible to diarise these items, or to get them done in between other tasks at work, or make time for them when you come home. If you have a spare moment and you know what to do, you won't just let it slip by doing nothing.

This is not to say you shouldn't make time to relax... but just sitting around doing nothing is entirely different. Even worse, you may have lots to do, and then have some spare time to do things, but not know where to start. You'll end up wasting your precious time and having to find more time to do all those things that could have been done earlier.

Use **power tools** whenever possible. For instance, use an electric drill or a saw to make your work easier, instead of struggling with tools that take more of your time. You can ignore this if it's just a small job that you don't do on a regular basis.

This also applies to **getting others** in who may be able to do things better or much quicker than you can, releasing you to spend your time on what you can do best.

Have a simple **filing system** at home. For this, you could have a **Master Plan** covering the important aspects of your life. You'll find an example of how to organise your Master Plan over the page, which will give you some ideas. When you have an up-to-date filing system, you will know where to put your letters and documents and not have to search for them when you need them.

Have only one filing system.

Please note that this may not all be for you. Take from it what you need or add your own variations.

Example of how to organise your Master Plan

A. EMERGENCIES

Telephone numbers and references for required documents should be located in your filing system.

B. PERSONAL FINANCES

1. Bank account

2. Credit/debit cards

3. Personal loans

4. Budget

5. Savings

6. Mortgage

7. Financial plan

C. INSURANCE

1. Life

2. Other

3. Last Will and Testament

D. AUTOMOBILE

E. TELEPHONE AND TV

F. HOME AND GARDEN

1. Maintenance

2. Groceries and cleaning etc

3. Gas and electricity

G. CHILDREN

1. Education

2. Activities etc

H. HEALTH

I.   RECREATION

J. SPIRITUAL AND INTELLECTUAL GROWTH

K. FAMILY AND FRIENDSHIPS

L.   MY MOST SIGNIFICANT RELATIONSHIP

For this Master Plan, you can use any file or case with dividers or into which you can put dividers with headings. When that folder is full, you can move half of it into a new one, repeating the process as required, but try to get rid of things no longer important/necessary when you put new ones in.

Include a duplicate of the full index at the front of each file for easy reference. If you have a PC, you can always add new sub-titles and print out the edited/updated index for each file when needed.

Some items may fill up a whole file, like the B file for your finances. Ensure that every important piece of paper is in its place within the right section. File correspondence that doesn't need immediate action in the correct place as soon as you open your letters. If you really don't have time, mark the paper with the letter and number of the section where it should be filed later. Make sure you do file it later!

If you have an existing filing system that you're happy with, keep it and keep using it. If your files become congested, try taking out a couple of sheets of paper that are no longer in use for each one you add. Then you won't need to buy another file for a long time.

Consider utilising a box file where you can place **every piece of paper** that comes into your home that you don't have time to respond to or file immediately. Then, you won't waste time looking for misplaced bills, receipts, etc. Empty this file regularly and file items in their right place, once a week at the very least. **It must not become a storage place for loose papers that are never filed.**

Have a "**transit file**" or **"action file"** that you take with you when you leave home; a particular folder in your briefcase, say, or in the bag you always take to work. This is for important things that need to be done that day. Remember to look at it when you get to work and when you get home. Even if you work from or in the home, getting into the habit of using an action file can make life easier. Reserve a place, like a small table or shelf near the front door, for your action file along with your incoming and outgoing mail.

Have a separate file for your **goals and projects**. This is a very special file, the contents of which should be revised regularly and with excitement! This will enhance your motivation to accomplish tasks. Remember that it mustn't be dormant. Transfer the actions that need to be carried out into your diary.

See this area as a very important component of your **inner manager.** These goals should bring direction to your life. If you do not have your own goals, others will fill your time with their goals.

In your Goals and Projects file, enter a few of the following, (not too many), and write them down as: **short-term** (weeks or months); **intermediate** (years); and **long-term** (more than 5 years) **projects**. For instance: getting the garden/house in order (short-term); planning and saving for a holiday or a study project (intermediate); planning and saving for a bigger home or retirement (long-term). There may be many others, depending on your circumstances.

Have the following in place for each project:

1. Specific project (definition).

2. Expected end result (in detail).

3. Steps to follow (break down into smaller sections and add tasks for the day, as a slot becomes available).

4. Availability of time and money to do it (be specific, make a budget and set deadlines).

To accomplish your goals, you need to take action. Write this down in your diary with the other things you need to do. Get a trusted friend or partner to go through your diary regularly with you.

Make notes of the following in your diary:

1. Upcoming events and appointments, as well as important tasks and regular tasks.

2. Try to fit in something from the list of tasks you should have completed.

3. Try to fit in some aspects of your projects.

Just like your transit file, **carry a journal** with you wherever you go. Jot down ideas and things to do as they come to mind. When you get home, tear out the page and deal with it as you would your other papers, e.g. file it in your file system or action file.

Collect a library, which may just be a few valuable books to be read regularly to help you achieve you goals. **Make a photo library** for your memories. Remember, a picture speaks a thousand words.

The experience of how good it feels to know where everything is, and that you are not disorganised and falling behind in doing things on time will be your reward. Also remember that no one is perfect, but that you are taking steps in the right direction. This knowledge alone makes it a satisfying undertaking.

You may already have many of the things I've mentioned in place, or you may have an existing system that works well for you. That's fine. If not, **don't be discouraged**. Take one step at a time, and though it may take a while for you to become organised, you will get there if you persist. Hopefully you will become much more organised than you are now!

Last, but not least, have a notebook near your telephone and regular work place. Never write messages or telephone numbers on loose pieces of paper. If you are not near your work place, write messages down in your "portable diary".

Using the checklists

Over the page, you'll find a checklist. The first one is quite long, so you might find it a little daunting. Start by just reading it through. As you read it, think how the points fit in with your own life.

To start the ball rolling, let's take the first point on the checklist: **a place for everything.**

You will have to choose if you want to do this, and come to a firm decision before you start.

You will need to think about it first, especially if this is a problem in your home. Discuss it with other members of the family (or write things down on paper if you live alone), and make a plan of action. Then go to the next point of getting rid of the things you don't need, storing others and putting the things you want to use regularly in their right places.

This process will in itself require **thinking, saying and doing**. If you apply this same thinking, saying and doing process with the items on all the checklists, you'll find that your life will start to get more organised and you will become much better at prioritising and doing the important things that will really make a difference. Taking the first step is often the hardest but it's worth the effort.

Getting Organised Checklist **Think, Say and Do**
❖ A place for everything.
❖ Everything in its place.
❖ Don't walk empty-handed (Make this a habit).
❖ Have a notebook in place for messages and telephone numbers.
❖ Get and fill in a main diary and calendar.
❖ Get and fill in a portable diary.
❖ Plan by day, week, month, and year for tasks and projects.
❖ Schedule time for leisure activities.
❖ Be ready for emergencies.
❖ List everything still to be done. Eliminate one by one.
❖ Prioritise P1, P2... NP1, NP2... L1, L2...
❖ Have your Master Plan and filing system ready.
❖ Get your action/transit file working.
❖ Plan your goals, projects and resolutions – short-term, intermediate and long-term. (Add chunks of projects in with your daily tasks.)

Getting Organised Checklist **Think, Say and Do**
❖ Use power tools and get others to help who can do the job better or quicker.
❖ Use a journal (Full pages go into your filing system).
❖ Read books to help you achieve.
❖ Make a photo library – your memories.
❖ For each hour you work, spend at least five minutes reading and organising.
❖ Make time for regular, important tasks, e.g. watering, cleaning, washing, cooking etc.

Remember, for each item on the checklist:

1. **Think**, imagine, and 'see' a clear picture.

2. **Say**, discuss and write down.

3. **Do,** practise and continually evaluate.

Either at the beginning, during the process, or before you carry out a task, come to a definite **decision** to do it, and see yourself accomplishing it.

HELP FOR ADULTS

Overcoming poor concentration

My advice for adults is virtually the same as I give to teenagers; so don't be surprised to see a lot of the same things here if you are reading both sections.

Under most circumstances, your ability to concentrate on any one thing for a length of time may be poor. There are exceptions, such as when you are very interested or excited about something, but in general, this is mostly not the case.

It's not only your ability to concentrate, but also the **quality** of your concentration that may affect your memory. There may also be a deficit in switching back to the initial activity when your attention was distracted to something else.

For instance, if you go to do something, and something else gets your attention along the way, you may forget what you wanted to do in the first place. You can imagine how dangerous this can be when driving, even if you just take a little longer to switch back after your attention was caught by something beside the road.

"Looping back" your attention means making an effort to keep contact with your initial activity. If your mind wanders off onto something else, it shouldn't become a separate "circle", but just a "loop". This habit, which you can teach yourself, may involve going back to the original point where you were distracted.

For example, if you are walking to get something, you may forget along the way what it was that you wanted until you go back to where you started. Use this in other situations; go back to the moment or circumstance when the thought of what you wanted to do came into your mind. When you arrive at the place where you were going to do or fetch something, there may be all sorts of other things competing for your attention. Always ask yourself, "What did I come here for in the first place?" and do that first.

Try and concentrate on one thing at a time. You should also keep in mind what you are going to do next, so you don't waste time between activities and interrupt your flow of concentration.

Always have another activity "up your sleeve", then if you get stuck with the first one for a while you won't waste time while waiting. Don't have too many things on your table when you work, ideally only those necessary to carry out the task at hand.

Try to work with as few external distractions as possible, e.g. check emails only at certain times, and don't allow other people's urgencies to interrupt you if you are in the position to do so. (You might have to be careful how you communicate this to your boss!)

Have everything ready before you start anything. Looking for things is not only a waste of time, but also a big distraction.

You may also have "internal distractions" over which you have less control. These may be due to tiredness or emotional issues. For people with "Inattentive" ADHD or ADD, attention deficit may frequently be internal as well as external.

Daydreaming may be an important component of internal distractions. Of course, to daydream at work is not acceptable, but to have dreams about the future is a good thing.

A word of caution: if you find that you "switch off" at times and lose contact with your environment, or frequently get an overpowering need to go to sleep, this should be investigated further, and you should talk to your doctor about it.

Try not to bring home issues to work and vice versa. If you are in the habit of doing private things during work hours, don't be surprised if this is the reason why you struggle to get your work finished properly on time.

Take regular exercise (if you work sitting down, try to go for a walk at lunchtime), try to get enough sleep, and stick to a healthy diet. Give yourself "a reward" for a task well done. This may differ according to the size of the task, from just getting up and walking around for a while, to an evening out, or a few days holiday. Plan bigger rewards in advance as this will help you to finish your tasks within the allocated time limit.

Give all your attention to one thing at a time. Let this be the most important thing you can do at that time.

Be active and productive, but don't allow yourself to be overloaded. It's easy to take on other people's responsibilities if you accept them. Learn to say "no" graciously.

Overcoming Poor Concentration Checklist

Think, Say and Do

❖ Split bigger tasks into smaller manageable ones, but not so small that you lose momentum.

❖ Create bigger chunks of time for finishing important tasks.

❖ Follow a specific program and a steady flow of activities.

❖ Keep just the papers you need for the current task on your desk.

❖ Eliminate external and internal distractions as far as possible.

❖ Follow a healthy exercise, eating and sleeping plan.

❖ Reward yourself for completed tasks. Don't work all the time.

❖ Concentrate on one task at a time, but keep the next one in mind so you don't lose continuity.

❖ Always "keep another task up your sleeve" to carry on with if you get temporarily stuck on the first one.

❖ Get everything ready before you start anything.

Remember, for each item on the checklist:

1. **Think**, imagine, and 'see' a clear picture.

2. **Say**, discuss, write down or draw pictures about.

3. **Do,** practise and continually evaluate.

HELP FOR ADULTS

Developing your memory power

Forgetfulness may be one of the signs of ADHD, but there are many people who don't suffer from ADHD who forget things just as easily. This may be one of your most annoying problems, and is usually not understood by those with a good memory.

A person suffering from ADHD may have just as poor a short-term or "working" memory as someone who suffers from Alzheimer's disease. This may bring criticism from others, and resentment towards them from you. Don't allow this to happen. There are positive steps you can take to fix this.

You may try memory training exercises which are freely available online. The problem is remembering to apply them! Don't let it get you down. Have various reminder systems in place, like writing things you want to remember in your notebook or diary. Always put your car keys in the same place, like your coat pocket, so you can drive off without spending half an hour looking for them. Make it part of your daily routine to put everything in the right place as soon as you get home or get to work.

Look regularly at your diary, and enter reminder items on your mobile phone or other device as well, if you have one. Use your memory for everyday activities, but don't rely on it for important appointments and things that have to be done.

There are some practical steps you can take to help make remembering easier. Here are a few suggestions:

❖ **Use a proper diary**. This may be electronic (PC or mobile phone calendar) or a journal-type book. Try to put everything important in this for later reference. I also use a pocket-sized diary with a small inside pouch on the front and back covers in which I slip my credit card and driver's licence and I always keep it with me. (You may of course carry these in your wallet) Note down the things that you want to transfer to your main diary, or things you want to remember when you don't have your main diary with you. Keep a small space available for your priorities for the day, which you will delete when they are completed, or transfer to the next day if unfinished. This method works for me, but do it the way you find best.

❖ **Always make a list when you go shopping.** This helps you determine exactly what you need to buy, and saves you from wasting money.

❖ **Have other reminders in place, apart from writing them in your diary.** Use your mobile phone, or electronic organiser for this – something you always carry with you.

❖ **Have a specific place for things to go, and never leave them elsewhere.** For example, if you put your keys somewhere other than their regular place when you come into the house, you may feel sure you'll remember where you put them. But, after a while, the memory's gone, so you'll waste time searching for them.

❖ **Get into the habit of doing certain activities at a regular time.** Watering the plants in the lounge on Sundays, for example. Creating simple habits like this will help you to remember.

Developing Memory Power Checklist
Think, Say and Do

❖ When you walk, take a keen interest in the things around you. Even make a note of some, and see how many you can remember later. Make it a habit each time you go for a walk.

❖ Associate people's names and faces with others you already know, or with other characteristics.

❖ Create funny and strange mental images to make remembering easy and fun.

❖ Try to relax by taking deep breaths, (not hyperventilating!), and relax the muscles you don't actively use at that point – you will concentrate better.

❖ Put keys, diary etc. in the same place. Never anywhere else.

❖ Make use of your mobile phone for reminders.

❖ Get into the habit of doing things regularly at a certain time.

Remember, for each item on the list:

1. **Think**: "plan in mind".

2. **Say**: discuss, write down, and draw pictures.

3. **Do**: practise, and evaluate.

www.adhd123.net

Driving with ADHD

Even if you take just a little longer to switch back after your attention is briefly caught by something beside the road, a second or two can mean the difference between good driving and an accident. Because you have ADHD, you need to work very hard to stay concentrated when driving.

People with ADHD are up to four times more likely than other drivers to be involved in motor vehicle accidents, and to get 12 points on their licence. There are suggestions that people with severe ADHD should only be allowed to drive when they are on medication. This is not law yet, but it's even more reason to be extra careful when driving.

Try not to do things impulsively, like suddenly turning into a side road without indicating in advance. If you don't see a turn until the last minute, go past and find a place to turn around. Always try to relax and concentrate on the road ahead.

Never talk on a mobile phone, or get so absorbed in a conversation (or argument) with your passengers that you forget what is going on around you. If you are new driver, it is a good idea to limit your passengers for the first year or so (see page 33). Don't get distracted by external things on the side of the road.

Make a habit of putting your seat belt on before you drive off, and not along the way. Stay within the speed limit, even if others may get impatient – that is their problem, not yours! Be even more careful not to drink or take drugs when you drive. The police will take it very seriously if you do something wrong, and your breath smells of alcohol.

HELP FOR ADULTS

Hearing, sight and coordination

A number of people with symptoms of ADHD have problems with coordination. This is can be a real problem at work, where you may be embarrassed by your clumsiness. Try to avoid difficult situations wherever possible, and try to laugh at yourself rather than feel humiliated.

You may think you can't see or hear well. However, if your ears are fine, and wearing glasses doesn't improve your sight, you have to find ways to compensate, like looking and listening more intently, or avoiding activities where good vision and hearing are vital.

If you tend to get lost when driving, you may need to invest in a satellite navigation system, follow a good map, or have someone travel with you to give you directions. Fortunately, like anything else, once you have followed a route several times, it will become easier to remember.

Developing and Maintaining Momentum

The earth spins due to its momentum, and we are travelling with it at a speed varying from zero at the poles (just rotating) to more than 1000 miles per hour if we are near the equator. Time is ticking away consistently related to the earth's momentum.

If we walk, our momentum is in a forward direction, as long as we do not stop or turn sideways.

When I was a student, I was working with an ambulance crew in the countryside. One day when we were travelling, the left rear wheel came loose and, as we ground to a halt, I saw the wheel running away from the ambulance with one of the male nurses starting to run after it. The wheel had momentum when on the ambulance, and when it came loose it kept its momentum for a while. I saw it hit an ant hill and jump in the air, and when it landed on the other side it carried on running until eventually it lost its momentum, as it had no driving force of its own. **The secret is** for us to cultivate our inner **driving force** by **steady attainment,** to keep our **momentum**. That is an important part of our executive function.

Definition of momentum: (*Encarta World English Dictionary*)

1. **Capacity for progressive development:** the power to increase or develop at an ever-growing pace.
   - eg *"The project was in danger of losing momentum."*

2. **Forward movement:** the speed or force of forward movement of an object.
 - *The momentum gained on the downhill stretches of the course.*

I do not suggest that we must be like a wheel driven by a motor, but that we should have momentum in ourselves, which will be "spinning quietly" even when we are relaxing, ready for the next action. Keeping momentum going in your actions is part of your winning strategy. Now this is easier said than done, especially for the person with executive function problems as I elaborated on in the middle section of this book, but there are ways to tackle it.

1. Make quality decisions about what you want for your life and write them down in detail in your journal. Keep looking at them and make additions and adjustments as time goes on. (This should be your "life's momentum") Be excited about it and don't be put off from pursuing it by anyone or anything. Get help if you feel you can't do it alone.

2. Plan your day and try to stick to it, but don't get frustrated about obstacles or interruptions, see them as unavoidable but temporary things to overcome; keep your momentum going. Live above the circumstances and not under them. If things get very difficult, just say to yourself: "I will just be the best me I can be." Don't give up, but be prepared to make adjustments when changes seem necessary. If you have not much to do on a specific day, make sure you do something, even if it is just one small thing. The planning, the doing and thinking about it afterwards, is enough to get you out of bed!

3. Concentrate on one thing at a time but have the next in mind so you do not have to waste time between tasks thinking what to do next.

4. Motivation and goal setting are key factors in keeping momentum. Also: Discipline. Commitment, Attitude, Fitness

(both physically and mentally), Integrity and the Dedication of time for important tasks or projects. (See the chapter on organizing; it will help you to stay in control.)

Here is a simple mathematical formula to help you:

KNOWLEDGE (PLANNING) (5) + EFFORT (5) + TIME NEEDED (5) + ATTITUDE (RELATIONSHIPS) (5) CONSISTANCY (5) = RESULT (25)

You can use this formula to evaluate the outcome of any task by giving points for the result. Then look at the items on the left side of the equation, and decide on which items you are losing points. Try to improve on these.

5. Always reduce the size of any task by taking away visible chunks. Just as an example: If the kitchen is in a mess, first put away or discard everything that doesn't need to be on the worktops. Then tackle the dishwasher if you have one, afterwards wash the bigger stuff and do a final clean up. With every chunk removed the task will look less daunting and will surely add to your momentum. Even if you are called away at some stage, you will know exactly where to recommence. Apply this strategy to any sizable task you have to do.

6. If you have a problem of being on time for meetings or getting things done in time, try the following: Plan the steps you have to follow "backwards." For instance, if you need to attend a meeting at a specific place at **10 am** and you have to go by train, you may estimate that the venue is a 20-minute walk from the station, so you will have to arrive there no later than 9.40. If the train journey takes 30 minutes, you need to be on the train as near to 9 o'clock as the timetable allows. If it takes

you 15 minutes to get from your home to the station, add on another 15 minutes for parking and buying a ticket, and don't leave your home any later than **8.30 am**.

Another example might be that you have to finish a writing project at a certain time (**19th November**). Start by estimating how long it will take to edit your final draft (7 working days – 10th November), how long it will take you to write the draft (14 days – 25th October), how much time you need to gather information about it (14 days – 11th October), how long to write down the details of the project (2 or 3 days – 7th October), so the project must be started in the first week of October. You don't have to be rigid, but see the different dates as deadlines. There are simpler and quicker ways to plan, but this one will work for anyone who needs a strategy.

Momentum – keeping going – is dependent on discipline rather than feelings, though feelings may affect it. The speed that you walk, for instance, may be affected by your mood but, even so, if you keep putting one foot in front of the other, you will still cover the distance. This is not to say that you must live your life by some rigid framework; a healthy self-discipline allows for some flexibility – but not an excessive amount.

The two greatest enemies of keeping a steady momentum are distraction and procrastination, both of which drag you down. You now know how to deal with these "enemies". Go through the checklist that follows, to see how to improve your momentum, which is a most important aspect of your life.

Developing and Maintaining Momentum Checklist
Think, Say and Do
❖ Have both long and short term goals.
❖ Plan to reach those goals.
❖ Implement the "Formula for Success".
❖ Take away one chunk at a time from a task.
❖ Be on time for meetings and deadlines.
❖ Control distractions e.g. phone calls, time on the PC, TV.
❖ Trying to follow a healthy diet and get enough sleep, exercise and recreation.
❖ Stay on one task until it is finished, even if you work intermittently because of impulsivity and inconsistency.
❖ Concentrate on one thing at a time, but have the next in mind.
❖ Try not to make sudden decisions. "Sleep on it" first, and discuss it. But when you do come to a final decision – go for it!
❖ If the task is not that interesting, keep your mind on the reward.

❖ Adhere to a definitive program; doing at least the things that may have important positive or negative consequences. If you agree to this, have a reward and punishment system in place.

❖ Don't be afraid to admit if you've made an incorrect decision due to impulsiveness and put it right.

❖ Cultivate a healthy self-discipline so you are not ruled by your feelings.

Remember, for each item on the list:

1. **Think**: "plan in mind".

2. **Say**: discuss, write down, and draw pictures.

3. **Do**: practise, and evaluate.

If you keep on struggling with this aspect of your life, you may want to add these books to your library (you should really have one, even if is a shelf in a cupboard):

Delivered from Distraction – E.M. Hallowell and J.J. Ratley

How to get things done *without trying too hard.* – R. Templar

Money Management

Never spend more than you earn – it is not yours. The easiest way to monitor this is to draw four columns on graph paper each month with the following headings: **expendable income; total expenses for month; accumulated income for year so far; accumulated expenditure for year so far.**

Instead of wishing your income were higher, scrutinise your present expenditure. Is there any way you can save on your mortgage, gas, electricity, telephone, groceries, eating out, etc.? Here are some suggestions to help curb overactive spending:

❖ **Keep good records of your expenditure.** "A place for every pound, and every pound in its place." Budget your money – even to the last penny or cent.

❖ **Pay and save.** Pay every debt and have a definite savings plan for worthwhile things. Keep someone informed about your finances, (like a partner or friend), and talk to them before you embark on any significant spending. It is of the utmost importance to contain impulsive spending. Remember that if you buy something new it will be second-hand immediately after you've bought it and worth much less than you paid for it.

❖ **Try to live on 80% of your income from your first pay cheque onwards.** Save 10% for the day you retire (this will grow to a significant amount with interest). Give the other 10% to the church, a charity organisation or to individuals who are less fortunate than you. This is just a suggestion, but most people acknowledge that giving your time and money to others in need brings more happiness into their own lives.

(Jim Rohn suggests in one of his presentations that we should actually try to live on 70% of our income and use the third 10% to release our entrepreneur spirit, e.g. by buying something, repairing it and selling it for a profit.)

❖ **Use self-discipline.** If you're a homeowner, and you sell your home to buy another one, try to discipline yourself, and put the profit you make on the first house into the next instead of spending it on something else. Otherwise you will always stay at the bottom of the "property ladder".

❖ Pay your taxes on time.

❖ Don't invest money you can't afford to lose.

❖ Pay any bill that has to be paid ASAP. You will have one less worry.

These guidelines will probably not address all your financial problems, but they will go a fair way towards doing so, and will keep you out of many troubles. If you are already in trouble, get professional help before it goes any further. Don't give up; even if it is sometimes painful, there is always a way out.

Money Management Checklist
Think, Say and Do

❖ Put income and expenditure controls in place (cash flow).

❖ Scrutinise expenditure.

❖ Plan for a better income (improve qualifications etc).

❖ Get out of debt – make a repayment plan.

❖ Save for worthwhile things, rather than buying cheap ones.

❖ Make a retirement plan.

❖ Keep someone trustworthy informed about your financial situation.

❖ Pay bills on time, including tax.

❖ Have an emergency fund for unforeseen expenses.

❖ File all financial documents and correspondence.

❖ Have a place to keep important shopping receipts.

Remember, for each item on the list:

1. **Think**: "plan in mind".
2. **Say**: discuss, write down, and draw pictures.
3. **Do**: practise, and evaluate.

HELP FOR ADULTS
Healthy relationships

This is such an important issue, possibly the number one reason for **happiness** and **unhappiness** in this world. (The second being financial problems and mismanagement.)

In short, relationship problems stem from the following:

- ❖ Misunderstanding the feelings and intentions of others;

- ❖ Being dishonest or inconsistent;

- ❖ Doing things without sharing them with those who ought to know, (e.g. spending money without telling your partner);

- ❖ Not doing what you said you would do, breaking promises.

Again there is no quick fix for these issues. You just have to learn from experience and observations, and on the way you will make mistakes. Everybody makes mistakes in their relationships. Never see yourself as a failure, but instead learn from it and try not to make the same mistake again.

Here are few suggestions to help you.

Realise that communicating with others is the essence of life, whether you like it or not. When you are in someone's company, give them your full attention. Look at their face and body language, and listen to what they are saying – to the words and the meaning behind the words. For some, it may be more difficult, but you will gain experience if you try. Only reply after you have listened well, and don't expect others to know what you want to say if you don't say it. They aren't mind readers!

Be honest in your dealings with others. Don't say what you don't mean, and on the other hand, don't say things that will hurt or annoy others if it is not absolutely necessary – even if it is true. If it must be said, try not to say it at a time that will be embarrassing to the other person, (i.e. in front of other peers). **You need friends not enemies.** If you have a reason to praise someone, do so, but with honesty and not flattery. Try to get into the habit of not talking about others if they are not there, as it may be construed as gossip. If the conversation turns in this direction, don't participate; it is very easy to be drawn into gossip.

Sometimes it is very difficult to tell the truth, especially if you have been told off many times in the past. But it is always better to tell the truth, and take the flack, rather than to lie about something. The truth has a funny way of coming out in the end, and lies will become even more of an embarrassment to you. They can also get you into big trouble.

Integrity goes with honesty. This means to be true, and true to yourself, at all times. It also means to **take responsibility** for your own actions, and not to blame others all the time. If you hit someone's car in an accident, or even just scratch it slightly, take responsibility. Don't look around to check whether or not someone saw you, and drive away. That's breaking the law.

The same factors apply to relationships. If you "scratch" someone, at least apologise and try to make amends for the damage. If the other person is still upset, then let it remain their problem, not yours. The most important step you can take on your road to being a person with integrity is to keep your word (promises) every time. Even if it hurts – in the end it will repay you.

As a child, you may have been told off and rejected frequently and this made you feel angry. The people who are with you now may not know where you come from, and may find it difficult to understand why you become so easily angered, leading to more rejection and unhappiness. Early attachment problems are sometimes overemphasised, but they are real and affect the person's sensitivity to rejection. In this respect, attachment and rejection may be two sides of the same coin.

There is professional help available. Ask your GP to refer you to a psychiatrist or a psychologist. Also think about issues that may have hurt you in the past and discuss them with someone you can trust. Talking with someone helps clarify things in your own mind.

Try to leave your past behind you, and to turn to a new page. Don't expect more from others than you expect from yourself, and remember – forgiving brings even more healing to you than to the other person. Keep "short accounts" of what others do to you. If you have said or done something that affects a relationship, saying that you are truly sorry about it seldom fails to bring healing, even if it takes some time for the other person to forgive entirely.

Do not qualify your apology by telling the other person what he or she did. If necessary, discuss this before you come to the point of apologising.

Try to see the good in others and respect them for that, even if it may not be easy. If you are in a position to help someone, just do it –you will find it greatly rewarding.

Do not intrude on other people's personal space. This means not coming too near or being too familiar with people you are not intimate with. This doesn't mean that you shouldn't have warmth and love for others – that is something different. It is a matter of having a healthy respect for yourself, and for others – including what is their property. Have you noticed how much effort we put into **borrowing** something we need, and how little effort we put into **taking it back** when we don't need it any longer?

Stand firm when everything goes wrong around you, even if you are accused of something you haven't done, or maybe even if you are to blame. Don't lash out and try to protect yourself by attacking others. Rather, see how the problem can be solved, and do the best you can in the circumstances. This is not to say you should allow others to "walk all over you". Stick to your principles, and try not to over-react.

Try to surround yourself with positive people who will pull you up instead of pulling you down. Sometimes this will mean that you actively have to seek the right friends and, unfortunately, also learn to spend less time with others. The exception is if you have a definite plan of action in mind to try and "pull someone up", without allowing the opposite to happen.

I have no doubt that a stabile and nurturing family is still the best place in which any child, with or without ADHD, can grow up. Give it to your children. Remember that we never own them nor do we own any other person; they are lent to us to look after and to help them to be the best they can be.

To help you with your attitude towards others, ponder on this little poem:

Let me live in a house by the side of the road,

Where the race of men go by,

Of men who are good, and men who are bad,

As good and as bad as I,

I will not sit in the scorner's seat,

Or hurl the cynic's ban,

Let me live in a house by the side of the road,

And be a friend to man

A verse from the poem written by **Sam Walter Foss**

There is a checklist over the page.

Relationships Checklist
Think, Say, and Do
❖ Try to be genuinely interested in others.
❖ Do things together with other people.
❖ Have your own view, but allow others to have theirs as well.
❖ Learn "body language." Read about it, and apply the knowledge.
❖ Do what you say you will do, and keep your promises.
❖ Discuss things with people you trust.
❖ Try to be honest and consistent in your dealing with others.
❖ Admit when you are wrong. Don't go around loaded with guilt.
❖ Look after yourself well (be clean and neat).
❖ Be fair to yourself as well as to others.
❖ Deal with anger. You may need help for this. Your GP can refer you for anger management if it's needed.
❖ Learn not to intrude into other's personal space, verbally or physically.

Remember: Think: plan in mind. **Say**: discuss, write down, and draw pictures. **Do**: practise and evaluate.

HELP FOR ADULTS

Overcoming feelings of failure

If I can really say, I've been the best me I can be, no guilt should come my way.

Don't be afraid to live! Don't give up on life, even if you feel like it. Don't stand back and let life pass you by! None of us can say that we never experience feelings of failure at times. This is a part of life that helps us to grow. The question is, "How do you respond to these feelings?"

There is a difference between failing to accomplish something and feeling like a failure – in the same way as there is a difference between doing a bad thing on occasion and being a bad person.

I know you have heard this before, but you need to view apparent failures as challenges. Learn from them, adjust, and do something in a different way if necessary.

Attempt to get it right from the beginning. How? Make sure your ladder is against the right wall before you climb it! You won't see a sunset if you run east – but if you are convinced that you are running west, go for it! At some stage, you will see the sunset. Don't keep looking back over your shoulder because you will never see it there.

There is an old saying, "Measure twice and cut once" which is very true. Another is, "Water the plants and not the weeds". I accept that it's often difficult not to water both, but in life this means putting emphasis on the good, and not on the bad.

If a banana won't peel from the base, try to peel it from the tip. If you struggle with plastic coverings that seem to get stronger as you grow older, always have a small pair of scissors with you!

Look after your belongings and they will "look after" you. These include your money, your work, your house, your car, your pets, and everything you own. Even your spouse and your children are given to you to care for, though you are not their owner. Look after other people's belongings with as much care and respect as you do your own, or more.

If you take on new responsibilities or tasks, do them without neglecting the regular ones. For instance, don't get so involved in a particular task that the plants in your garden die because you didn't water them. At least you should make sure that somebody else does it for you when you can't, or invest in a watering system.

There are about 6 billion people on Earth, but amongst all these people, **there is only one person exactly like you**. **This makes you very special.** Only you have your eyes, your ears, hands, feet, brain etc. You may or may not like yourself, but you have to come to terms with it. This is who you are, and what you have, so you might as well use it to the best of your ability.

Tell yourself today, **"I am going to be the best me I can be."** Not better or worse than others, just the **best version of yourself**. Then see what a difference this makes in your life. Be grateful for what you have, look after it well and use it for yourself and others. That is all you need to and can do – but be sure to do it.

There is a saying, "What you don't use, you lose". Turning this idea around, "use what you have and it will get better".

Apply the principle to other areas of your life; say to yourself today:

"I will be the happiest me I can be."

"I will be the most efficient me I can be."

Say it over and over with excitement, and see what a difference it makes. Come to the point where you say with all your heart, "I only want to be me!" You may include your name as well.

Remember that many important people in history either suffered from ADHD, or had significant features of the condition. Two names that come to mind are Einstein and Winston Churchill. If you study their lives, you will recognise the symptoms.

Because of their "scattered brains" people with ADHD have more "lateral" thoughts, and may stumble on ideas other people never thought of. If they can only hold onto and pursue these ideas, they can deliver creative solutions and great innovation. Often, we feel at our most creative when the pressure is on. But you have to train yourself to do things not only when the pressure is on at the last minute, but consistently. This will not come easily by itself, and this is why you have to practise the principles outlined in this book.

Having said all this, don't be frustrated if, in spite of all your best efforts, you are still making mistakes. If you suffer from the symptoms of ADHD, you will have an especially hard time doing things consistently right. Don't get discouraged, or feel guilty about it. If you tried your best, this is all you can humanly do.

This also applies to your expectation of others. Say to yourself every day:

"This is the first day of the rest of my life, and I am going to make it my best day."

Learn from the past, but don't dwell on your failures. When you make a mistake, see if there is anything you can learn and do about it – then do it differently next time.

Ask yourself, "How long am I going to feel bad about this?"

Then answer yourself with, "Maybe 10 minutes, or half an hour!"

For this time, feel bad; then tell yourself, "Stop!"

This usually works better than continuing to feel guilty about something you can't do anything about.

There is one thing worse than being poor, and that is not to have a plan. And there is one thing worse than making a mistake, and that is to do nothing.

If you do well today, it will look after tomorrow and, though you have to plan, worrying won't help you. Always be on your guard. However, don't be sloppy or irresponsible. The fact that you have symptoms of ADHD should never be used as an excuse to do wrong or bad things to others.

One thing you have to come to terms with in life is what you can do and what you can't do, even if you do your very best. Only life experience can teach you this. Don't keep on bumping your head against the wall; there will always be a door.

Overcoming feelings of failure
Think, Say, and Do

❖ Think of areas in your life that you need to improve, and do something about them.

❖ Is there space for more discipline in your life? Think of specific areas that need attention.

❖ Think of past "failures", and how you can learn from them.

❖ Don't neglect regular responsibilities when taking on new ones.

❖ If something goes wrong, try to correct it as best you can, feel bad about it for a short time, and then stop!

❖ Listen to criticism, but ignore it if it's not true.

❖ Cultivate the habit of making the best of every situation.

❖ Be enthusiastic without constantly criticising yourself or others.

❖ Put more effort into doing what you can do best.

❖ Wherever possible, mix with positive people who will build you up. Develop the habit of not joining in with gossip.

❖ Come to terms with things that you are not able to do.

Remember: Think: plan in mind. **Say:** discuss, write down, and draw pictures. **Do**: practise and evaluate.

Bringing it all together

The principles in this book have worked for others, **and they can work for you.** But you are the only person who can take the responsibility to make them happen in your life. You must **act** on what you know and believe in. Remember the saying:

"What you don't use, you lose"

And remember the flip side:

"Use what you have and it will get better"

Don't carry unnecessary things with you, or drag unfinished tasks behind you. Finish (lose) one task before you tackle the next. It will make your "journey" so much easier if you don't have to worry about the things you haven't done.

All that I have presented here is given to you with my sincere desire to help you on your way to a happy, productive life. Take hold of it with both hands and build on it. Make it your own. Don't blame yourself if you fail. Attempt to do it differently and better next time. You just need to be diligent in everything you do.

"The plans of the diligent lead to profit as surely as haste leads to poverty."

(Proverbs 21:5)

Final Checklist	Yes	No	When?
❖ Have you got your emergency file in place? If you are away and the car breaks down, will the people at home know exactly where to find the number to dial? What about your life insurance policies? Are they just between other papers somewhere?			
❖ Have you got your finances in order? If you buy something today, will it be added to the expenses side of your budget?			
❖ Are you happy with what you eat? Have you begun to eat some fruit every day, maybe starting with an apple?			
❖ Have you started with an exercise program yet?			
❖ Are you regularly doing something about the things in your action file?			
❖ If you have children, are you spending at least 10 minutes quality time with each of them, where they take the lead in what you are doing, or plan something special together?			

Final Checklist	Yes	No	When?
❖ Are you spending quality times with your partner/friend? Or making that telephone call you wanted to make for a long time?			
❖ Do you stay within the speed limit, even if the driver behind you gets impatient?			
❖ Are you exercising your authority where it is needed? Are your children looking at programs or doing things you know they shouldn't because you are following the path of least resistance? Are they going to bed on time?			

Using this checklist

Write Y or N, depending on whether your answer is mainly No or mainly Yes. Use the When column to write the date when you achieved something or the date when you are going to do something.

Useful resources

These are a few selected organisations that I recommend, based on my personal and professional experiences, where you can get more information about ADHD and other commonly related problems. The list is very selective and biased toward the UK, but you can find others that are local to you by searching on the Internet.

Information About ADHD

ADDISS (National Attention Deficit Information and Support Service)
www.addiss.co.uk

ADDERS (An online ADD/ADHD information service)
www.adders.org

ADDMIRE (ADD Multi-Agency Information Resource)
www.ADDmire.org.uk
(This is the local website for Ashford and St Peter's Hospitals, where I work)

Adult ADHD/ADD

www.netdoctor.co.uk/adhd/adult/

www.adhd123.net

www.adhdcoachonline.com

Marital and Partnership Problems

Specialised advice: www.adhdmarriage.com
Relate: www.relate.org.uk

Even though you may or may not have any problems with your marriage or partner relationship just now, you cannot go wrong in enrolling in The Marriage Course, which is internationally acclaimed and is available at many local venues.

The Marriage Course: www.relationshipcentral.org

Family Relations

Dr James Dobson has done great work for families and has written numerous articles and books on family relationships.

www.focusonthefamily.com

Drug and Alcohol Dependency

www.talktofrank.com

I would also recommend that you talk to your GP as soon as possible, as they should be able to recommend the best sources of help and support in your area and refer you to expert help for your underlying problems.

Advice on Debts

The following UK organisations all offer advice with debts and financial problems.

Citizens Advice Bureau
Website: www.citizensadvice.org.uk
(Or visit your local office)

National Debt Line
Website: www.nationaldebtline.co.uk
Freephone: 08088 084000

Credit Action
Website: www.creditaction.org.uk
Tel: 01522 699777

A final word

Some of the thoughts expressed here are my own. Others come from lectures that I've listened to and books I've read, not necessarily on the subject of ADHD. I want to mention just a few of my sources that come to mind: Jim Rohn, Jeff Davidson, Anthony Robbins, David Allen, Vic Conant, some information from the book of Christopher Green, and the many speakers I have listened to on the subject of ADHD over the years. This book is a compilation of ideas and I have attempted to put together as many of them as possible. If even one of these ideas helps you, it will make me happy.

I have taken the liberty not to give a detailed bibliography, as this book is not intended to be a scientific dissertation of the subject, only about practical things that work.

"For when the One Great Scorer comes to write against your name,
He marks – not that you won or lost – But how you played the game."
Grantland Rice

Don't be afraid to *live* your life

If you have any comments about this book, please feel free to drop me an email at info@adhd123.net.

Made in the USA
Middletown, DE
17 May 2017